HOLLAND MATHEMATICS STUDIES 138

ND – AMSTERDAM • NEW YORK • OXFORD • TOKYO

TH

JÁNOS BOLYAI

APPENDIX

The Theory of Space

With Introduction, Comments, and Addenda

Edited by

Prof. Ferenc Kárteszi
Doctor of the Mathematical Sciences

Supplement by
Prof. Barna Szénássy

1987

NORTH-HOLLAND – AMSTERDAM • NEW YORK • OXFORD • TOKYO

ISBN: 0 444 86528 4

Publishers:

ELSEVIER SCIENCE PUBLISHERS B.V.
P.O. BOX 1991
1000 BZ AMSTERDAM
THE NETHERLANDS

and
AKADÉMIAI KIADÓ, HUNGARY

Sole distributors for the U.S.A. and Canada:

ELSEVIER SCIENCE PUBLISHING COMPANY, INC.
52 VANDERBILT AVENUE
NEW YORK, N.Y. 10017
U.S.A.

For the East European countries, Korean People's Republic, Cuba, People's Republic of Vietnam and Mongolia:

Kultura Hungarian Foreign Trading Co., P.O. Box 149,
H–1389 Budapest, Hungary

PRINTED IN HUNGARY

CONTENTS

Part IV

The work of Bolyai as reflected by subsequent investigations 177

Supplement 220

(by Barna Szénássy)

PREFACE TO THE ENGLISH EDITION

This book is a revised edition of the memorial volume I wrote in 1952, by invitation, to the 150th anniversary of János Bolyai's birth. That time, I could spend only two months with writing the text and drawing the illustrations. Therefore in the second edition I have somewhat revised and corrected the original.

Encouraged by people abroad interested in the subject, I gave consent to publish my book in English. However, for the better information of these readers I stipulated that the book should be supplemented with a brief historical survey. The task was taken on by Professor Barna Szénássy. Using the latest documents, he wrote a concise historical supplement. I believe that learning some facts of Hungarian political and science history will help the less informed reader get acquainted with the miserable fate of János Bolyai and with his intellectual world.

Initially, at several suggestions, I thought that the book written a quarter of a century ago should be completely renewed and made more conforming to the demands and style of today. In fact, recent efforts have more and more aimed at a definitive showdown of the intuitive elements of knowledge still to be found. The excessive freedom of traditional scientific style and language should be eliminated through the systematic use of a modern, strictly formalized language. This is a remarkable point of view, indeed. Accordingly, I ought to present non-Euclidean geometry in the most up-to-date manner and comment on Bolyai's work in that connection.

Doing so, however, I could not make evident what an epoch-making discovery the geometry of János Bolyai was in its own time; it should be emphasized how natural Bolyai's ideas, his revolutionary aspect of mathematical space theory seem to be today and what it has since grown into. On the other hand, the very thing the reader interested in the history of science shall clearly see is that contemporaries, excepting perhaps Gauss and Lobachevsky, were averse to Bolyai's thoughts and considered them an artificial, or even obscure, intellectual construction.

For these reasons, the commentator of a classical work that has strongly influenced the development of science must present his subject by putting it back to its own time. He must sketch the antecedents as well as describe the difficulties rooted in the relative primitiveness of contemporary scientific opinion and hindering the evolution of new ideas. Also the triumphal spreading of the new ideas, especially in its initial period, should be described. I tried to write my book so as to serve this purpose.

In our country, Bolyai's work is generally called the *Appendix*. In Bolyai's Latin manuscript the title *Scientia Spatii* (*Raumlehre* in the German one) occurs. For the sake of historical fidelity, I gave my book the title *Appendix, the science of space*. To bring out the historical point of view, the facsimile of a copy of Bolyai's work printed in June 1831 is also included, though I have prepared — in cooperation with György Hajós and Imre Trencsényi Waldapfel, late professor of classical philology — a careful translation of the text into modern Hungarian and added it to the Latin original.

Part I is a historical introduction which makes use of the latest literary sources. The Supplement, written for the English edition by Professor Barna Szénássy, completes this part and helps the foreign reader.

Part II contains the Latin original and its translation into present-day language. Though the translation follows the requirements of modern language and style, it accurately reflects the concise Latin text. Dissection into chapters not occurring in the original, changing notation to that used today, setting the illustrations at suitable places of the text, application of an up-to-date drawing technique and, finally, the presence of some new illustrations help to avoid the unnecessary difficulties usually encountered when reading old prints and texts.

Part III is a series of informal short remarks divided into sections corresponding to those of the original work. Actually, with these remarks we try to make easier the comprehension of the text which, because of its conciseness, can only be read with close attention, thinking the material over and over again. This happens once by completion, once by reformulation and more detailed explanation in order to dissolve conciseness, yet another time by addition and further argument.

In Part IV, by picking out and sketching some important topics, we attempt to indicate the effect that may be traced in the development of modern mathematics after János Bolyai's space theory had become generally known.

Ferenc Kárteszi

PART I

EVOLUTION OF THE SPACE CONCEPT UP TO THE DISCOVERY OF NON-EUCLIDEAN GEOMETRY

1. FROM THE EMPIRICAL STUDY OF SPACE TO DEDUCTIVE GEOMETRY

As far as we nowadays know, in pre-Greek times a great deal of empirical knowledge had already accumulated, and this collection of practical facts served for Greek genius as a source in creating the deductive science called geometry. We also know that speculative logic had initially developed independently of mathematics and reached a high level before its application to empirical facts concerning space began. From that time onward, it was not only logic which assisted the development of geometry, but geometry has also reacted on the evolution of logic. In pre-Greek mathematics the concepts of theorem, proof, definition, axiom and postulate had not yet occurred; all of them are creations of Greek intellect.

According to the history of science it was THALES (624–548 B. C.) who, on the visual level, began to arrange the facts gained by experience and to search for explanations which reduce the complicated to the simple ("demonstration" in a "perceptible" way); the evolution in that direction started in his times.

One century later the application of the methods of speculative logic for proving geometrical assertions was begun. In this period (450—325 B. C.) the following circumstances deserve special attention. From the very beginning, extremely rigorous, exact proofs were produced. The method of indirect proof was used remarkably often. The validity of many geometrical statements which had been known and obvious for a long time was proved.

The first textbook of geometry, entitled *Elements,* was written by HIPPOCRATES (about 450—430 B. C.), who attempted to put contemporary geometrical knowledge in a strict logical order. Hence one concludes that geometry began to turn into a deductive science in the period before HIPPOCRATES.

The book of EUCLID (about 325 B. C.), also entitled *Elements*, partly rests on former works and is a synthesis of deductive geometry, as created by the Greeks, in a perfect system (here the word "perfect" refers to the level of science reached at that time, and not to demands emerging in the course of later progress).

EUCLID's work is a textbook in the best sense of the word. It teaches us what kind of requirements should be raised against scientific knowledge, in which way facts should be treated, and how to pass on the results obtained. All this is done in the highly purified manner which had evolved as a fruit of long-lasting meditations of the Greeks.

Elements attains these aims indirectly, by providing a model. It arranges the material in groups: definitions, postulates and axioms are coming first, succeeded by the statements and proofs of the theorems already known (actually, for the most part, known for a very long time).

EUCLID himself does not at all explain what the point of this grouping is. He may have assumed that the intelligent reader would find out the motive of the scheme by getting acquainted with the work in its entirety and thinking it over and over again. Probably the fundamental principles of the order revealed in the book had already crystallised and become current in science to such an extent that they required an exact and possibly complete realisation rather than a mere exposition.

We know that Greek scholars, as early as in the days of PLATON, had recognized the following: in the chain of mathematical proofs there is no "regressus ad infinitum": mathematics must have foundations which cannot be proved any longer. The formation of these principles had required more than a century of immense intellectual effort. In the possession of these principles, Euclid could start composing his work from a high level of scientific thought.

Elements begins with 35 definitions, 5 postulates, and 5 axioms. They together have been called *foundations* (principles), for EUCLID endeavoured to deduce all the rest of his work by a logical process starting from these foundations and relying only on them.

It should be noted that in *Elements* definition, postulate and axiom mean something else than they mean today. We will not make a linguistic analysis or an appraisal of the foundations by modern standards. Instead, we are going to state the foundations in an up-to-date language.

We cite the first three definitions for clarifying the sense, different from present-day usage, of the word "definition", and the last one.

1. Point is the thing that has no parts.
2. Line is length without breadth.
3. The ends of a line are points.

----

35. Those straight lines are parallel which are in one plane and which, produced to any length on both sides, do not meet.

The postulates are the following.

1. A straight line may be drawn from any point to any other point.
2. The straight line may be produced to any length.
3. Around any point as a centre, a circle of any radius may be described.
4. Any two right angles are equal.
5. If a straight line meets two other straight lines so as to make the sum of the two interior angles on one side of it less than two right angles, then the other straight lines,

if produced indefinitely, will meet on that side on which the angle sum is less than two right angles.

The axioms read as follows.

1. Things which are equal to the same thing are also equal to one another.
2. Adding equals to equals the sums are equal.
3. Subtracting equals from equals the remainders are equal.
4. Things which can be interchanged (can replace one another) are equal.
5. The whole is greater than its part.

If the reader of *Elements* looks at the shortcomings of the foundations in possession of present knowledge and from the viewpoint of today, if he projects this aspect into the past and applies it to the appraisal of the book, to weighing its significance, he will be at a loss for understanding the irresistible influence exerted by EUCLID's work on contemporaries as well as on scholars of two thousand years thereafter. If, however, we consider the ideas of those scholars active in the times of EUCLID or later on who have built their theories on unmotivated, contradictory, vague and wrong foundations and tried to base the whole of science on one or two principles, we understand the immense success of the book and its lasting effect on scientific throught.

EUCLID's inheritance should be studied free from false interpretations and added distortions of fetishists and successors who have considered it a dogma. This is not an easy task. Euclid's program, his intentions, are to be made out from the original contents of his work.

In this way it seems likely that EUCLID did not aim only at a formal demonstration of the theorems, but also investigated their necessary interdependence or, at least, made an effort to clear up the relationship between them. There is no other possibility to understand why he has given an apparently complicated proof of the following theorem:

An exterior angle of a triangle is greater than either of the angles of the triangle that are not adjacent to it.

Although from Postulate 5 it follows easily that the sum of the angles of a triangle equals two right angles which yields the theorem in question, Euclid did not choose this proof. Obviously, he realised that the validity of Postulate 5 is not a necessary condition for the theorem to hold.

According to later terminology, theorems independent of Postulate 5 are called *absolute* theorems. EUCLID tried to find the absolute theorems and proved them (similarly to the example cited above) as consequences of the foundations needed. He has stated neither his program nor the outcomes of this discovery attained during the realisation. Nevertheless, the arrangement of the theorems and the careful selection of the proofs attest that he saw the significance of the fundamental hypothesis on parallel lines more clearly than his contemporaries and had it figure almost unobjectionably in the construction of geometry. We say "almost unobjectionably" since he derives, for instance, the *transitivity* of parallelism ($a \| b$ and $b \| c$ imply $a \| c$), an absolute theorem, from Postulate 5.

Further investigation of absolute theorems as well as the complication of a perfect system of foundations, striven for but not achieved by EUCLID, have been left to posterity.

2. ATTEMPTS TO PROVE POSTULATE 5

In a subsequent but also very old *Euclidean* text Postulate 5 appears as Axiom 11. In the course of time the distinction between postulate and axiom faded away, and nowadays one prefers to speak of axioms and their system. The conviction arose that the treatment of geometry should start from possibly few statements of simple content relying on visual perception and not subjected to any proof; these are called axioms forming the system of axioms. If experience and visual perception are not referred to any more, and all theorems are deduced by means of logical arguments based on the axioms, then the treatment is said to be *axiomatic*. EUCLID's Postulate 5 is called also the *axiom of parallelism*.

For brevity, by the *residual system of axioms* we shall mean the set of all Euclidean axioms excepting the axiom of parallelism. If some theorem A can be deduced from the Euclidean system of axioms, and if the statement of the axiom of parallelism can be deduced from the statements of A and the residual system of axioms, then proposition A may be used as axiom instead of the axiom of parallelism. It may happen that such a theorem which can replace the axiom of parallelism seems simpler than the original axiom. A possible substitute for the axiom of parallelism says that the sum of the angles of any triangle is equal to two right angles.

A proposition's being "simpler" than another one is a subjective statement, we cannot even formulate exactly what it means. Nevertheless, the first critics of EUCLID have already striven for simplifications of this kind, and the search for substitute axioms has, up to the beginning of the 19th century, revived again and again. Although the nice axioms did not advance geometry, we are going to cite a few of them, since they deserve some attention from the didactic point of view.

1. In the plane, we consider a straight line and a point not on this line; among the straight lines of the plane through the point there is exactly one that does not intersect the given line (PROCLUS, 5th century A. D.).
2. In a half-plane bounded by a straight line all points lying at the same distance from the line form a straight line (CLAVIO, 1574).
3. There exist two similar but non-congruent triangles (WALLIS, 1663).
4. For any triangle, there exists a triangle of greater area (GAUSS, 1799).
5. Three points lie on either a straight line or a circle (FARKAS BOLYAI, 1851).

EUCLID deduced from the residual system of axioms that, in the plane, to a straight line through a point not on the line *at least one* parallel line may be drawn. However, he could not deduce from the residual system of axioms that there is *only one* such parallel line. By the evidence of the cautious proofs he gave for the theorems deducible from the residual system of axioms it seems quite probable that he has recognized or guessed the truth: the theorem "... exactly one parallel line may be drawn" is not an absolute theorem, that is, it cannot be deduced from the residual system of axioms.

There were two possibilities available: to complete the residual system of axioms with either the theorem "... exactly one ..." as an axiom (not deducible from the system) or some equivalent axiom. EUCLID chose the latter way. The fifth axiom has served this purpose.

The critics of EUCLID's work have not understood that this was a necessary procedure and, in fact, EUCLID's greatest achievement. Later on, the fifth axiom was even regarded as the only blot on Euclid's ingenious work. Efforts have been made to remove this blot and prove that the statement of the axiom of parallelism could be deduced from the residual system of axioms. However, all these efforts failed.

After EUCLID, up to the initial spreading of KANT's philosophy, 55 works have dealt with the problem concerning the axiom of parallelism in Euclidean geometry, and 67 new works appeared between 1760 and 1800. This fact supports the point of view adopted by the history of science that in the above period of fourty years the rapid increase of the interest in a question open for two thousand years was caused by the controversies about Kant's philosophy.

Here we pick out just two examples from the extensive literature and merely sketch them: the tentative proofs of GEROLAMO SACCHERI and JOHANN HEINRICH LAMBERT which were published in 1733 and 1786, respectively, will be discussed. The basic idea of both of these trials was to start from the set consisting of the residual system of axioms and the negation of Euclid's fifth axiom and arrive, through some argument, at a contradiction, which would prove the validity of the Euclidean axiom. In the course of long, sharp-witted reasoning they did find startling theorems, but no intrinsic logical contradiction. Because of their philosophical views, however, they were unable to notice, or unwilling to admit, the failure.

SACCHERI, in his work entitled *Euclides ab omni naevo vindicatus: sive conatus geometricus quo stabiliuntur prima ipsa universae Geometria Principia* (Milan, 1733), gave essentially the following tentative proof *(Fig. 1)*.

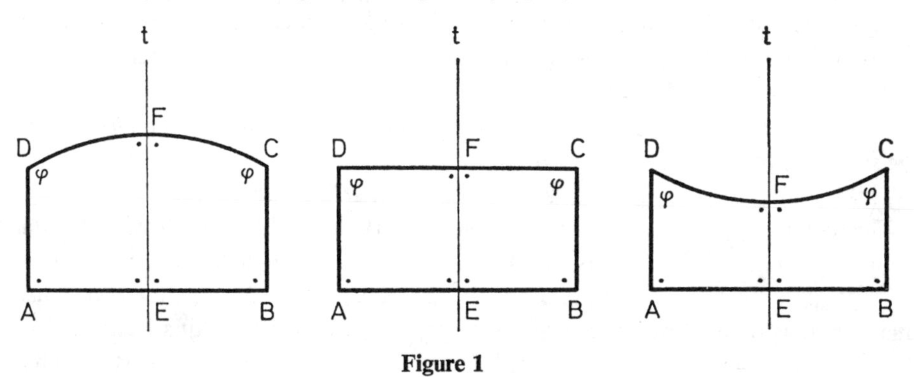

Figure 1

Let the sides AD and BC of the quadrangle $ABCD$ be equal to each other, and let the angles at the vertices A and B be right angles. Is it possible to determine, with the help of the residual system of axioms, the size of the angles at vertices C and D of the quadrangle? By reflection in the line t, the perpendicular bisector of the base AB, it easily follows that t perpendicularly bisects the side CD too, and that the angles in question are equal. Denote these angles by φ, and the right angle by ϱ. It should be decided which of the hypotheses

$$\varphi > \varrho, \quad \varphi = \varrho, \quad \varphi < \varrho$$

on the angle φ is implied by the residual system of axioms. The quadrangle $ABCD$ defined above is called *Saccheri's quadrangle.*

SACCHERI has easily disproved the first one, the *hypothesis of the obtuse angle.* The *hypothesis of the right angle* is equivalent to the Euclidean axiom of parallelism. If one succeeded in disproving the *hypothesis of the acute angle,* the validity of $\varphi = \varrho$ would be established, that is, the Euclidean axiom of parallelism would turn out to be a consequence, obtainable by logical means, of the residual system of axioms.

SACCHERI claimed to have disproved also the hypothesis of the acute angle, but this is not the case. Through a deduction starting from the hypothesis $\varphi < \varrho$ he actually arrived at a theorem which seemed to be false and he there finished the chain of ideas, aiming at a disproof, by the statement "... and this contradicts the nature of the straight line"; the theorem, however, contradicted a property of the line that follows from the hypothesis $\varphi = \varrho$.

Thus SACCHERI has only proved that if $\varphi = \varrho$ is true, then $\varphi < \varrho$ is not, an assertion requiring no proof. Though he has reached the door to the discovery of non-Euclidean geometry deducible from the hypothesis of the acute angle, he got stopped outside, since stiff Euclidean aspect and philosophical views closed the door before him just after it had opened a little.

Among the several lemmas appearing and correctly proved in Saccheri's work there is one to which we turn special attention. Its importance would become clear if we tried to decide experimentally whether the properties of physical space are more truly reflected by Euclidean or by non-Euclidean geometry based on the hypothesis $\varphi < \varrho$. Later the two BOLYAI's were considering the idea of such an experiment, whereas LOBACHEVSKY and GAUSS carried it out. The form of the theorem reminds us of WALLIS' axiom cited above:

If one of the relations $\varphi > \varrho$, $\varphi = \varrho$, $\varphi < \varrho$ holds for one Saccheri quadrangle, then the same relation is valid for all such quadrangles.

This theorem may rightly be referred to as *Saccheri's angle theorem.*

The tentative proof occurring in LAMBERT's work entitled *Theorie der Parallellinien* (Leipziger Archiv, 1786) begins with the study of a quadrangle having right angles at three of its vertices *(Lambert's quadrangle)*. Denote the fourth angle by φ. *Saccheri's* quadrangle is divided by its line of symmetry into two *Lambert* quadrangles. Three cases are possible: $\varphi > \varrho$, $\varphi = \varrho$, or $\varphi < \varrho$. Just as Saccheri, Lambert intends to prove that the first relation as well as the third one contradict the residual system of axioms. His deductions are simpler than, but essentially similar to, those of SACCHERI. Nevertheless, LAMBERT admitted that in the course of his argument starting from the assumption $\varphi<\varrho$ he could not arrive at a contradiction. He has still claimed Euclidean geometry to be unique possible. Thus he has already opened wide the door, but did not dare to enter the new world of non-Euclidean geometry.

Some remarks of LAMBERT went so deep that if he had not been frightened by his own results, he might have become the creator of a new theory of space. Let us sketch a few of them.

If α, β, γ are the angles of a triangle, then the hypothesis of the obtuse angle is equivalent to the relation

$$\alpha+\beta+\gamma-2\varrho=\varepsilon>0.$$

On the other hand, the hypothesis of the acute angle is equivalent to

$$2\varrho-\alpha-\beta-\gamma=\delta>0$$

(ε and δ are called the *excess* and *defect*). Therefore the assumptions on φ can be written, respectively, in the form

$$\varepsilon>0, \quad \varepsilon=\delta=0, \quad \delta>0.$$

For a triangle of area T, according as $\varepsilon>0$ or $\delta>0$, the equation

$$T=k^2\varepsilon \quad \text{or} \quad T=k^2\delta$$

holds, where the positive quantity k^2 is unknown but *has the same value* for all triangles. On the basis of these formulas Lambert conjectured that *the geometry built on the hypothesis $\delta>0$ was valid on the sphere with imaginary radius.*

This statement calls for explanation. If α, β, γ denote the radian measures of the angles of a spherical triangle on the sphere with radius r, then the surface of the triangle is $r^2(\alpha+\beta+\gamma-\pi)$ or, in short, $T=r^2\varepsilon$. If in the case of a triangle with defect we say that its excess is a negative number $(\varepsilon<0)$, then a mechanical modification of the area formula yields

$$T=k^2\delta=k^2(\pi-\alpha-\beta-\gamma)=$$
$$=-k^2(\alpha+\beta+\gamma-\pi)=-k^2\varepsilon.$$

This looks like the surface formula for spherical triangles provided that, led by formal analogy, we introduce not only the term *"triangle with negative excess"* but, in view of the negative number $-k^2$, also the notion of a sphere with radius $k\sqrt{-1}=r$, i.e., a *"sphere with imaginary radius"*.

This, however, remains a play of words as long as we cannot define the sphere with imaginary radius and the radius itself in a geometric manner, i.e., until the verity content of the word is unknown (one century later it was BELTRAMI who proved that there exists a geometrical figure through which the above words may be given a geometric meaning). Thus the latter remark of LAMBERT was merely a play of words.

Another, exact, remark of LAMBERT is based on the circumstance that, in both of the geometries which correspond to the hypotheses $\varepsilon > 0$ and $\delta > 0$, the constant appearing in the area formula is independent of the angle sum of the triangle; hence there exist no similar non-congruent triangles in these geometries (cf. the substitute axiom of WALLIS from 1663).

By inference starting from the area formula, Lambert also discovered that in these geometries there should exist a unit of a length which is not chosen arbitrarily but derived in a natural way, i.e., from the properties of the plane. For selecting the unit of length on the basis of geometrical properties, he argued as follows.

The proportionality of the area of a triangle to the defect, the inequalities $0 < \delta < 2\varrho$, and the assumption that there is a triangle with any fixed defect δ satisfying these inequalities imply the existence of an equilateral triangle (denote its side by d) having just the fixed δ as defect *(Fig. 2)*. Thus, for instance, the side d determined by the half of the right angle as defect may be chosen for the natural unit of length.

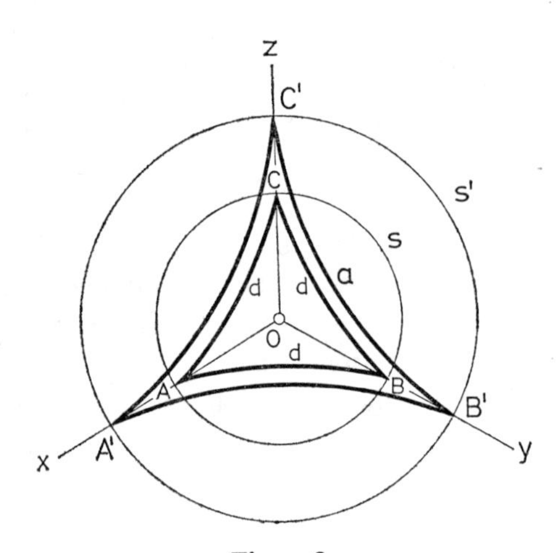

Figure 2

LAMBERT concluded this argument by the statement that the non-existence of a natural unit of length is equivalent to the negation of the hypothesis of the acute angle. That is why he said:

"This superb corollary creates the desire that the hypothesis of the acute angle be valid."

On the other hand, he declared rather definitely that the consequences resulting from the negation of the Euclidean axiom of parallelism cannot be accepted.

3. REVIVING INVESTIGATIONS AT THE BEGINNING OF THE 19TH CENTURY

More than a half century elapsed since P. STÄCKEL's book *The Geometric Investigations of Farkas Bolyai and János Bolyai* (Budapest, 1914; in Hungarian), written with great care and competence, had appeared. On p. 41 of this book one reads the following:

"The strong and lasting effect of Kant's 'Kritik der reinen Vernunft' (1781) revived the pursuit of the foundations of geometry and, in particular, the study of the theory of parallel lines."

Even contemporary philosophers knew that the theses of KANT's philosophy relating to the concept of space call for defence. They wanted to deduce the Euclidean axiom of parallelism from the residual system of axioms because in that way they thought to be able to verify the independence of geometry from experience. Even the considerations of GAUSS and FARKAS BOLYAI were initially influenced by this false opinion and only later they got free from the effect of Kantian ideas. Both of them reached a scientific conviction opposite to Kantian philosophy by research in geometry.

Out of the investigations published in the first third of the 19th century we are going to discuss only a few results of A. M. LEGENDRE (1752–1833), F. K. SCHWEIKART (1780–1855), and F. A. TAURINUS, and also an instructively wrong "proof" (published in Göttingen in 1818) due to B. F. THIBAUT.

Comparing the works of THIBAUT, LEGENDRE and several other authors with investigations of SACCHERI or LAMBERT an almost incomprehensible relapse may be noticed. Instead of proceeding towards the perception of the possibility of a new space concept, a direction in which LAMBERT had already started, they obstinately tried to establish by quite primitive and amazingly bad proofs that the Euclidean concept of space is unique possible.

By an argument based on the residual system of axioms, using the rotation of the plane around a point and the composition of three such rotations, THIBAUT "proved" that the sum of the exterior angles of a triangle is equal to four right angles and, consequently, the angle sum of the triangle equals two right angles. We describe the essence of his reasoning by the aid of *Fig. 3*.

Let the triangle ABC rotate about its vertex A to the position AB_1C_1 in such a way that the angle $C_1AC=\alpha^*$ is an exterior angle at A of the original triangle; hence A is an interior point of the segment BC_1. Similarly, turn AB_1C_1 about B so that the side A_2C_2 of the resulting triangle $A_2B_2C_2$ lies in the extension beyond B of the segment CB. The angle of rotation is now $ABC_2=\beta^*$. Finally, turn the triangle $A_2B_2C_2$ about the point C so that the side A_3C_3 of the triangle $A_3B_3C_3$ so obtained lies in the extension beyond C of side AC of the original triangle.

The three rotations result in a translation of length $b+a+c$ of the original triangle along the line AC in the direction of C. Altogether the directed line AC made a rota-

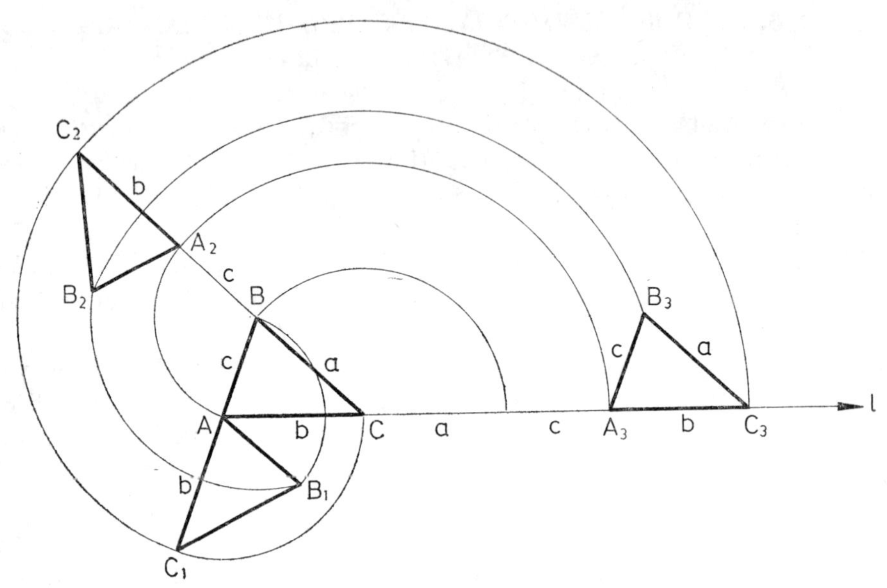

Figure 3

tion through the angle $\alpha^*+\beta^*+\gamma^*$ and got back to its initial position, i.e. turned by four right angles, which implies that the angle sum of the triangle is equal to two right angles.

This false proof roughly introduced the untrue statement that, owing to the residual system of axioms, two successive rotations by angles ω and τ yield a rotation of the plane by the angle $\omega+\tau$. The latter is true as an absolute theorem only if the centres of both rotations coincide. Otherwise the theorem is valid only in the Euclidean plane and is equivalent to the axiom of parallelism.

It would have been sufficient if THIBAUT had considered the composition of rotations of a sphere through angles ω and τ about two diameters, in which case he could have seen that the theorem does not hold under the hypothesis of the obtuse angle, and therefore it should not be accepted without criticism.

Indeed, let the regular spherical triangle ABC be a spherical octant, K its spherical centre, and X, Y, Z the midpoints of the sides *(Fig. 4)*. Rotation through a right angle about the point A carries the side $\widehat{AC}$ into $\widehat{AB}$, and rotation through a right angle about the point B carries $\widehat{AB}$ into $\widehat{CB}$. Both rotations yield the spherical motion $\widehat{AC}\to\widehat{CB}$. The single rotation through 120° about the point K results in the same motion. Thus in the present case $\omega=90°$ and $\tau=90°$, but the resultant motion, though a rotation about K, is through an angle of 120° rather than 180°.

Incoherencies similar to the false proofs of THIBAUT occurred in publications of such an excellent mathematician as LEGENDRE too. They have been noticed and criticised also by JÁNOS BOLYAI. The following remark appears in a manuscript of his written about 1834:

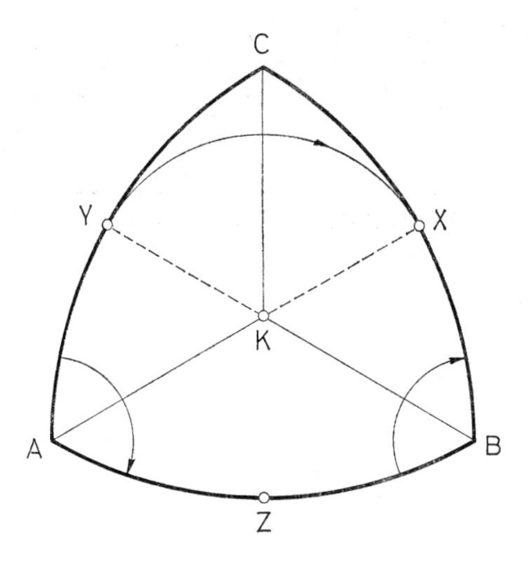

Figure 4

"The great Legendre, besides a previous wrong proof at which we must be surprised to have arisen in such an ingenious brain, ..."

LOBACHEVSKY has intensively dealt with LEGENDRE's geometrical publications. He dissected the defective proofs, pointed to the roots of the mistakes and, on the other hand, picked out the small number of valuable details where LEGENDRE had given a correct and coruscating treatment. Two of the latter, known as LEGENDRE's first and second angle theorems, will be given below (properly speaking, they belong to SACCHERI, but have become generally known under a different name).

The proof of the first theorem will be developed by reproducing, in essence, LEGENDRE's publication which appeared in 1800. In the bequest of GAUSS, on the cover of one of his books, the same proof may be read. Obviously, LEGENDRE's publication has escaped GAUSS' attention, since beside the proof he made the following note: "I found this on 8th November 1828". In the proof of the second angle theorem, however, we follow the reformulation given by LOBACHEVSKY.

Theorem. *The angle sum of a triangle is not greater than two right angles.*

It should be noted that LEGENDRE, as it is clearly shown by his proof, tacitly considered this theorem to be deducible from the residual system of axioms.

Proof. *(Fig. 5).* Suppose there exists a triangle ABC whose angle sum is greater than the sum 2ϱ of two right angles.

Starting from B, lay n further copies of this triangle on the line AB next to one another. Thus we have $n+1$ congruent triangles. It is an absolute theorem (proved as such by EUCLID) that the sum of two angles of a triangle is smaller than two right angles. Consequently, the angular domains α and β with common vertex A_k do not cover the half-plane: there is a gap of size $2\varrho-\alpha-\beta=\delta>0$ between them. As the

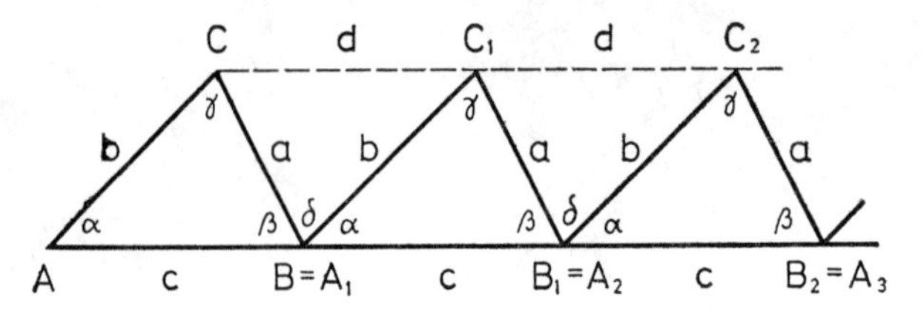
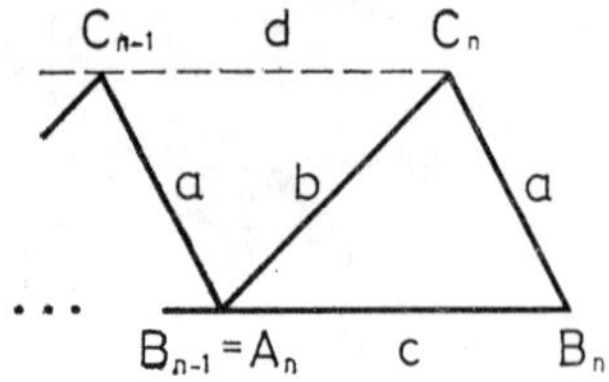

Figure 5

triangles CBC_1, $C_1B_1C_2$, ..., $C_{n-1}B_{n-1}C_n$ are determined by the sides a, b and the angle δ between them, they are congruent. Hence $CC_1=C_1C_2=\ldots=C_{n-1}C_n=$ $=d>0$.

Our assumption concerning the triangle ABC yields $2\varrho-\alpha-\beta<\gamma$, i.e., $\delta<\gamma$. The angle between the sides a, b in ABC and CBC_1 being γ and δ, respectively, we have $d<c$.

As a broken line which connects two points is longer than the connecting segment, the path $ACC_1C_2\ldots C_nA_n$ is longer than the segment AA_n. This can be expressed by the aid of the component segments as follows:

$$b+nd+b > nc.$$

Introducing the segment $e=c-d>0$ we obtain

$$2b > ne,$$

where n is any natural number. This inequality contradicts the axiom of ARCHIMEDES according to which $ne>2b$ if n is sufficiently large. The contradiction means that the starting assumption $\alpha+\beta+\gamma>2\varrho$ is not valid, i.e., $\alpha+\beta+\gamma\leqq 2\varrho$.

We note that the proof above deeply exploits the property of the straight line as to which if we lay segments equal to c on the half-line with endpoint A one after the other then the segments align without overlap however far we pursue (as we actually can) the process. In truth none of EUCLID's axioms, obtained by abstraction, perfectly expresses that property, though in his proofs EUCLID assumed the line to have it. The axiom of ARCHIMEDES has remedied this deficiency of the foundations (however, the complete system of the foundations, i.e., the system of axioms for Euclidean geometry was set up only in 1899).

Before stating the second angle theorem, we treat some lemmas. They occur in early works of LOBACHEVSKY, in connection with the analysis of LEGENDRE's proofs.

1. *If there exists a triangle whose angle sum equals two right angles, then there exists a Saccheri quadrangle each of whose angles is a right angle.*

For let ACD be the triangle mentioned in the lemma *(Fig 6)*. According to the assumption it has at least two acute angles, say the angles at vertices A and D. Then the perpendicular to AD through C intersects the segment AD at the point B located between A and D. The angle sum of the partial triangles ABC and BCD exceeds that of ACD by the angle 2ϱ arisen at B. Thus the angle sum of the two partial triangles is

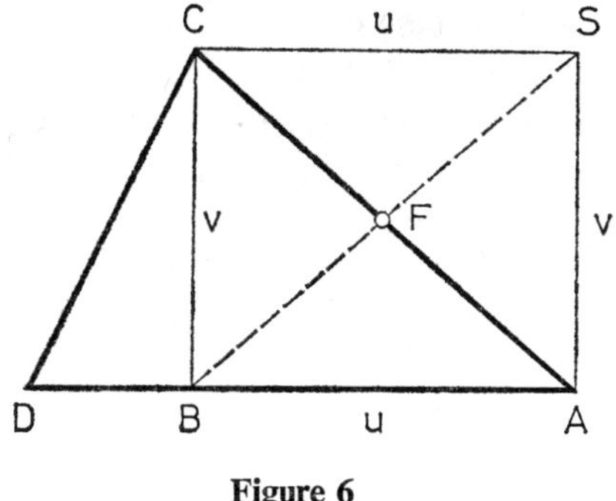

Figure 6

4ϱ. As the angle sum cannot be greater than 2ϱ in either of the two triangles, it is equal to 2ϱ in both of them. Reflecting the triangle ACB in the midpoint F of the segment AC we obtain the triangle CAS congruent to ACB. Obviously, $CBAS$ is a Saccheri quadrangle complying with the assertion of the lemma.

2. *If there is a Saccheri quadrangle with each angle a right angle and two neighbouring sides* **u** *and* **v**, *then there is one with neighbouring sides* **nu** *and* **nv** *too,* **n** *being any natural number (Fig. 7).*

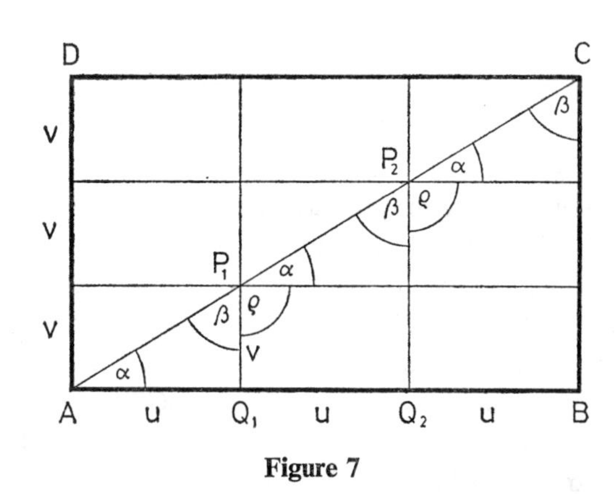

Figure 7

Since the angle and side measures of the special quadrangle with sides u, v appearing in the assumption have the same properties as those of a Euclidean rectangle (all angles are right angles and opposite sides are equal), the argument known from Euclidean geometry for the construction of a rectangle with sides nu, nv from rectangles with sides u, v may be applied to the case of this special rectangle.

3. *If the sum of both acute angles of the right triangle determined by the legs* **u**, **v** *is equal to a right angle, then the same is true for the right triangle determined by the legs* **nu**, **nv** *(Fig. 7).*

For the line AB passes through the points P_1, P_2, ..., i.e., $AP_1P_2...B$ is not actually a broken line, since $\alpha+\beta+\varrho=2\varrho$ by assumption. Therefore also the angles at vertices A and C of the right triangle determined by the legs $AB=nu$, $BC=nv$ are α and β, respectively.

4. *Any right triangle ABC may be covered by a triangle KLC whose angles are equal to those of a right triangle determined by the legs* **u**, **v** *and having the property described in Lemma 3 so that A lies between the points K and C of the segment KC, while B lies between the points L and C of the segment LC (Fig. 8).*

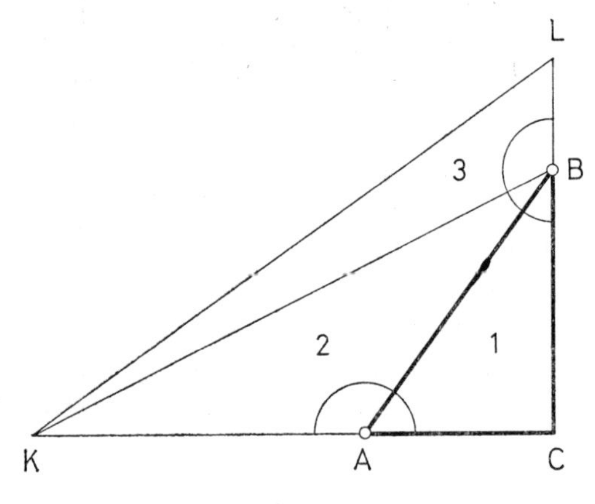

Figure 8

For let k and l be the smallest natural numbers with $ku>AC$ and $lv>BC$, and let $n>k, l$. The legs $KC=nu$ and $LC=nv$ belonging to this n determine a right triangle KLC the sum of whose acute angles is equal to a right angle. Since A and B are interior points of two different sides of the triangle KLC, segments AB and KL have no point in common. Therefore KLC actually contains ABC as a part.

5. *If the angle sum of the right triangle determined by the legs* **u**, **v** *equals two right angles, then the same is true for the angle sum of any right triangle.*

For let ABC be a right triangle *(Fig. 8)*, and let KLC be the right triangle of angle sum 2ϱ associated with ABC according to Lemma 4. The segments KB and AB split the triangle KLC into three triangles. Denote the angle sums and defects of these triangles by σ_1, σ_2, σ_3 and δ_1, δ_2, δ_3, respectively. As

$$(\sigma_1+\delta_1)+(\sigma_2+\delta_2)+(\sigma_3+\delta_3) = 6\varrho,$$

and $\sigma_1+\sigma_2+\sigma_3$ exceeds the angle sum of KLC by an angle of 4ϱ arising at the vertices A and B, we have

$$2\varrho+4\varrho+(\delta_1+\delta_2+\delta_3) = 6\varrho,$$

i.e.,

$$\delta_1+\delta_2+\delta_3 = 0.$$

By Legendre's first angle theorem $\delta_1\geqq 0$, $\delta_2\geqq 0$, $\delta_3\geqq 0$. Thus $\delta_1=\delta_2=\delta_3=0$. The relation $\delta_1=0$ is just what we had to prove.

In possession of these lemmas, LEGENDRE's second angle theorem is already easy to prove.

Theorem. *The angle sum of a triangle is either always equal to or always less than two right angles.*

Proof. If there is a triangle whose angle sum equals two right angles, then there is also a right triangle having this property, so that according to the lemmas any right triangle is of this kind. We proved in connection with *Fig 6.* that each triangle may be split into two triangles by a suitable altitude line and, consequently, has angle sum equal to two right angles.

If there is a triangle whose angle sum is less than two right angles, then all triangles share this property. In the opposite case there would exist a triangle with angle sum equal to two right angles (by the first angle theorem there is none with a greater angle sum). The existence of the latter triangle, however, yields a contradiction. Indeed, it implies that the angle sum of any triangle equals two right angles. In particular, this would hold for the triangle which we have assumed to be of angle sum less than two right angles.

LEGENDRE's second angle theorem was applied by GAUSS and LOBACHEVSKY for deciding the angle sum valid in physical space. It suffices to find, by accurate measuring, the angle sum of a single triangle with very long sides. If it is equal to two right angles, then Euclidean geometry reflects the nature of physical space exactly. If, however, measurement exhibits a defect, then Euclidean geometry describes physical space only approximately. GAUSS and LOBACHEVSKY have tried to determine the defect of certain triangles by geodesical and astronomical measurements, respectively. Their attempts failed. As to JÁNOS BOLYAI, he definitely declared that the problem could only be solved in an empirical way, perhaps by indirect conclusions drawn from experience concerning planetary motion rather than by optical measurement.

The views of LEGENDRE and his contemporaries were still far from the clear ideas of BOLYAI, GAUSS, and LOBACHEVSKY. Under the influence of KANT's philosophy, erudite people generally believed that the human brain could not even think of space having its structure different from that described by Euclidean geometry, and that the concept of space had not originated in experience.

The period ending with the investigations of LEGENDRE was followed by the awakening, however faint at the outset, of new scientific ideas. The life-works of SCHWEIKART and TAURINUS reflect a new spirit. The decline following LAMBERT's time is over. SCHWEIKART's results obtained between 1807 and 1818 are known from his short description prepared for GAUSS. In 1819, in a letter to GERLING, GAUSS wrote with appreciation about the ideas of SCHWEIKART. Later on, SCHWEIKART initiated also TAURINUS into his investigations and encouraged him to pursue them. TAURINUS developed his theory in the work entitled *Geometriae prima elementa* (Cologne, 1826).

SCHWEIKART's short description contains the following statements.

There are two kinds of geometry: Euclidean geometry and another one, in SCHWEI-

KART's words "eine astralitische Grössenlehre". In the latter kind of geometry the angle sum of a triangle is *not equal* to two right angles. Under this assumption it can be rigorously proved that

1. the angle sum of a triangle is less than two right angles;
2. the less this angle sum, the greater the area of the triangle;
3. if the congruent sides of an isosceles triangle are produced by equal lengths, then the length of the altitude from their common point tends to a finite limit.

In the case of an isosceles right triangle, let c denote the limit mentioned in 3. It is easy to prove that this "constant" is infinite if and only if the angle sum of a triangle equals two right angles, that is, if Euclidean geometry is valid.

On appreciating this description, GAUSS wrote (in a letter from the year 1819) that in this "astral geometry" all problems could be solved; thus the least upper bound of the area of a triangle was

$$\frac{c^2}{\{\lg(1+\sqrt{2})\}^2}.$$

It may be seen from neither SCHWEIKART's nor GAUSS' letter how clear ideas they have formed, to what extent they have worked out non-Euclidean geometry, and how near they have come to the level of perfectly exact treatment reached by BOLYAI.

The investigations of SCHWEIKART were carried on by TAURINUS, who elaborated the trigonometry arising from spherical trigonometry automatically when the radius r of the sphere is throughout replaced by $ir\,(i=\sqrt{-1})$. In this, analytically created trigonometry the property that the angle sum of a triangle is less than two right angles may be derived from the trigonometric theorems. TAURINUS, however, was unable to find out the geometric meaning of the "radius" ir playing a fundamental role in the analytic formulation. He regarded his theory as an interesting logical play (without a geometric interpretation of ir it is in fact a curiosity of that kind) and firmly believed in the exclusive reality of the Euclidean concept of space.

As we see from the history of the over two thousand years after EUCLID outlined above, no substantial progress has been made in the subject during that period. However, the erudite great mathematician GAUSS, who enjoyed numerous impulses through his wide personal contacts, began to see the contours of the emerging "new world" of geometry. For several decades, he made enduring efforts to establish non-Euclidean geometry for himself in a satisfactorily rigorous way.

It was the third decade of the 19th century that brought a decisive change. This was best expressed by FARKAS BOLYAI in a letter dated from 1823:

"Each idea has, so to say, its own epoch when it is discovered in different places at the same time, just as in spring the violets sprout wherever the sun shines."

4. THE MEDITATIONS OF GAUSS AND THEIR RESULTS

Up to now, there has not appeared any treatise that would examine the activity of GAUSS in elaborating non-Euclidean geometry and his reluctance to publish his own achievements or to make known those of others with the same rigour as that applied to the dissection of BOLYAI's and LOBACHEVSKY's works to the finest details. The value of the results of GAUSS is not reduced by the fact that he took some of his notes after considering informations received from others. All these questions might form an interesting subject for the history of science. In the sequel we content ourselves with sketching the facts certified by documents.

GAUSS knew the ideas and results of SACCHERI, LAMBERT, SCHWEIKART, TAURINUS, and FARKAS BOLYAI. Subsequently, he got acquainted with the work of JÁNOS BOLYAI (1832) and, later on, with that of LOBACHEVSKY (1840). He was aware of the geometric investigations of all contemporaries who dealt with the problem of parallelism in any noteworthy manner.

Concerning the role of GAUSS in the history of non-Euclidean geometry, DELONE writes on p. 93 of this book *The History of Science in the Soviet Union* (Akadémiai Kiadó, Budapest, 1950; in Hungarian):

"Gauss, as it is obvious from his works, thought over many things in this connection, but he published nothing;"

GAUSS did give a clear outline of his thoughts on the new geometry in confidential letters, but did not permit his ideas to be published. His plans, ideas and results have been reconstructed by the aid of the following sources:

a) His work entitled *Disquisitiones generales circa superficies curvas* (1827).

b) His letters written to WACHTLER (1816), GERLING (1819), TAURINUS (1824), BESSEL (1829), SCHUMACHER (1831), and FARKAS BOLYAI (1832).

c) His notes prepared for himself that were found after his death.

It can be seen from these sources that GAUSS has not sufficiently developed his ideas. In the systematic elaboration of the new geometry he was really "overtaken" by JÁNOS BOLYAI and by LOBACHEVSKY.

As early as in 1824 he writes to TAURINUS that from the assumption that the angle sum of the triangle is less than two right angles a logically correct geometry may be deduced where each of the problems known from ordinary geometry is solvable up to a constant which cannot be "a priori" determined. At the same place, he notes that the area of the triangle in this new geometry cannot be arbitrarily large.

He writes in 1829 for the first time that KANT's concept of space cannot be maintained. Subsequently, in a letter dated from 1832, he regards the geometry of JÁNOS BOLYAI as a brilliant evidence that KANT's assertion of space being merely a form of human understanding was wrong.

He puts down the formula (with proof) for the area of a triangle only in 1832. The proof is incomplete, since it states but does not establish that the area of the triangle whose sides are pairwise asymptotic straight lines is finite (although GAUSS notes that this assertion, too, must be proved).

Another significant formula reflecting, in concrete form, the results of his meditations occurs in his letter of 12th July 1831: the expression for the circumference of a circle in terms of the radius. In essence, he obtains the expression

$$\pi k\left(e^{\frac{r}{k}}-e^{-\frac{r}{k}}\right)$$

(k denotes the remarkable constant referred to in his previous letters). However, nothing is known of how he deduced it.

GAUSS knew from formulas of this kind that, for $k\to\infty$, the relations valid in non-Euclidean geometry turn into the corresponding relations of Euclidean geometry, and if in a formula the data of a figure are very small as compared with the unknown k (in our case, for instance, $r\to 0$), the statement expressed by the formula also turns into the proposition belonging to the Euclidean formula.

This discovery stimulated him to measure the angles of a triangle, large when considered under terrestrial circumstances, in order to determine the constant k valid in physical space. The same discovery made him search for the metric of that space enjoying the Euclidean property in an "infinitesimal" neighbourhood. Both of these attempts failed, but the statement of the problems is an evidence of the depth of GAUSS' ideas.

Comparing his letters and his notes written for himself it can be concluded beyond doubt that, up to 1832, GAUSS arrived at the following results.

He defined parallelism just as JÁNOS BOLYAI. He discovered everything that occurs in the ten first sections of BOLYAI's work.

He discovered that, in the plane satisfying the residual system of axioms, the tangent circle to a line a at a point A tends, as its radius increases ($r\to\infty$), to a limiting curve called *paracycle* three of whose points lie on a straight line if and only if Euclid's axiom of parallelism, too, is valid in the plane. Otherwise the paracycle is a curve that can be shifted in itself and that has for lines of symmetry every line intersecting the curve perpendicularly.

In the geometric system where the paracycle is a curve rather than a straight line, he found the expression for the area of a triangle as a function of a constant k and the defect of the triangle, and the expression for the circumference of a circle as a function of the same constant k and the radius of the circle. He gave a simple geometric interpretation of k. He pointed out that the value of k valid in reality can be obtained only by experience (measuring) but not deduced from the residual system of axioms.

Let us return to the most valuable, though defective, note found in the bequest of Gauss. The note, written probably between 1840 and 1846, is four pages long and gives a very concise sketch of the solution of a problem. GAUSS here tried to answer the following question: if the rules of Euclidean geometry are valid inside an infinitesimally small circle, what are the metrical relations in the large, that is, in the whole plane? In other words, he asked if there exist plane geometries, *Euclidean in the small,* which are different from the geometries of EUCLID and of BOLYAI–LOBACHEVSKY. He failed to notice, however, that another, stronger assumption had crept into his starting hypothesis: according to this tacit assumption, all motions in the plane taking a point into any other point and all rotations about any point are possible.

For brevity, call the explicit and the tacit assumption the *local* and the *kinematic* one, respectively. The investigations of GAUSS we are considering start actually from the aggregate of both assumptions. Only as late as in 1893 S. LIE's paper entitled *Theorie der Transformationsgruppen* revealed that the local assumption is a consequence of the kinematic one. So GAUSS solved, instead of the problem he had in mind, the following: *Taking the kinematic assumption in the plane for granted, what kind of geometry is valid in the plane?*

If we not only omit the axiom of parallelism from Euclidean foundations, but also replace the axioms of order by other ones, we can construct a different kind of non-Euclidean geometry (also *Gauss'* note under discussion refers to such a geometry). This may be derived from spherical geometry by regarding any pair of antipodal points of the sphere as a single point. Later on, this geometry got the name used nowadays: *Riemannian geometry* (in the strict sense).

We are now able to formulate the result GAUSS has achieved according to that remarkable note: *If in the plane the kinematic assumption holds, the plane is either a Euclidean plane, or a Bolyai–Lobachevsky plane, or a Riemannian plane.*

GAUSS, who has shown a wealth of creative activity on so large territories of mathematics and arrived at superb discoveries also in non-Euclidean geometry, may be called a discoverer but not a creator of non-Euclidean geometry. This opinion is expressed by the term *Bolyai–Lobachevsky* geometry which is widely accepted today.

5. THE GEOMETRIC INVESTIGATIONS OF LOBACHEVSKY

NIKOLAI IVANOVICH LOBACHEVSKY was awarded the title "magister of the mathematical sciences" in 1814. His inquiring mind and critical sense for the creation of mathematical concepts drifted him more and more definitely towards the solution of the problem of parallel lines challenging all thinkers of that time. In his case as well, the discovery of non-Euclidean geometry was the fruit of many years' research work governed by gradually clearer and clearer ideas. KAGAN says:

"It was not before the period 1826–29 that his ideas fully evolved. He realised their value only then."

In 1829 LOBACHEVSKY published the first account of his discovery in a long dissertation entitled *On the Foundations of Geometry*. When this work appeared, he was the rector of the university of Kazan. All the rest of his works on geometry revise and improve the material of that dissertation.

Contemporary mathematicians did not understand LOBACHEVSKY's works. That these works are hard to follow and the digestion of their contents requires a great deal of intellectual effort was mentioned also by GAUSS.

In 1840 LOBACHEVSKY, with the aim that finally his work would be comprehended and appreciated, wrote a new book on non-Euclidean geometry confined to an elementary treatment of the most substantial facts. The book was in German, 61 pages in small octavo size, and published in Berlin under the title *Geometrische Untersuchungen zur Theorie der Parallellinien.* It has been read by GAUSS and, eight years later, by JÁNOS BOLYAI, the two contemporaries who not only understood but also appreciated it. GAUSS began to learn Russian in order that he could study the other works of that "sharp-witted" mathematician as well.

GAUSS did not inform LOBACHEVSKY that BOLYAI had arrived at similar results, just as he had not informed BOLYAI of the geometric investigations of SCHWEIKART and TAURINUS. Thus it has never come to LOBACHEVSKY's knowledge that non-Euclidean geometry had been independently and simultaneously discovered by a Hungarian mathematician.

The philosophical views of LOBACHEVSKY can be characterised by a single quotation from his works:

"First concepts with which science begins are acquired through our senses; innate ideas are not to be believed in."

LOBACHEVSKY obtained his non-Euclidean geometry by a process different from that given in BOLYAI's work. He drew a comparison between Euclidean and hyperbolic (in his words *"imaginary"*, the term "fictive" being perhaps more accurate) geometries based on the assumptions

$$\alpha+\beta+\gamma = 2\varrho \quad \text{and} \quad \alpha+\beta+\gamma < 2\varrho,$$

respectively, concerning the angle sum of the triangle, and treated mainly those theorems whose formulations in the two systems are different.

He believed and tried to verify that in physical space the "imaginary" geometry was valid, and wanted to find the constant k characteristic for that geometry by astronomical measuring. In his last work *Pangeometry* (Kazan University Scientific Reports, 1855; in Russian) he got nearer to the spirit of BOLYAI's treatment directed towards a synthesis. In order to give an insight into his opinion on the connection between geometry and physics we quote the following sentence from his writings:

"Some forces of nature obey one and some of them the other, from our point of view specific, geometry."

It is a fact that LOBACHEVSKY's publications on geometry comprise, apart from laying the foundations of non-Euclidean geometry (that is, in the detailed and exhaustive discussion), significantly more than BOLYAI's work. Trigonometry, analytical geometry, differential geometry, and the methods of imaginary geometry applied to integral calculus are given a through and deep treatment in these publications.

The works of LOBACHEVSKY try the reader not only because of their length. Reading them meant hard work even to GAUSS as one sees from the following remark: "... these papers remind me of a dense forest through which it is difficult to find the way and the light ...". However, the booklet "Geometrische Untersuchungen ..." mentioned before is an exception. It is a pearl among introductory textbooks of mathematics ever written. GAUSS as well as BOLYAI called it a masterpiece.

LOBACHEVSKY, similarly to JÁNOS BOLYAI, could not live in his life the victory of his ingenious discovery.

6. THE MATHEMATICAL STUDIES OF JÁNOS BOLYAI

It cannot be our purpose to write a biography of JÁNOS BOLYAI. On the other hand, the history of the formation of non-Euclidean geometry necessarily includes a description of how the great problem, the suitable way towards its solution, the system of absolute emerged in his mind. On the basis of documents available, all this can be reconstructed.

FARKAS BOLYAI returned from Göttingen in 1799. In 1801 he got married. In 1804 he was invited to be professor of mathematics at the college of Marosvásárhely. Although the appointment offered modest income and meagre prospects, he accepted it with pleasure since at that time the welfare of his first-born son János and the health state of his neurotic wife compelled him to assure solid, however poor, material conditions. He spent 47 years in teaching and 5 others in retirement.

János was born in Kolozsvár on 15th December 1802. He spent considerable part of his childhood in the Marosvásárhely house got by his father as a payment in kind. Farkas devoted great care to the education of his son and gave János the first systematic instruction himself: he taught him mathematics, fencing, playing the violin, and the elements of music theory. János learned quickly and much; he made fast progress also in Latin. Farkas held that, at the age of 15, János could not learn anything more from him.

In the first period of his life, János was deeply impressed by his father. The mathematical interest of Farkas, his efforts at solving fundamental problems had a decisive influence on the ingenious child. It was also caused by the influence of his father that the interest in the problem concerning the axiom of parallelism, open for two thousand years, took strong root in him. Also his admiration and unlimited respect for GAUSS began in that time, following the stories related by his father.

Farkas wished that János pursued his promising mathematical studies with GAUSS. Because of the miserable financial circumstances of the Marosvásárhely professor, however, this could not be accomplished. Though János did not feel attracted by a career in the army, he continued his studies at the academy of military engineering in Vienna. This was the first great disappointment in his life. Really, also he would have liked to learn mathematics from GAUSS. He was admitted to the 4th class of the academy on 24th August 1818.

The academy did not afford János many novelties in mathematics. The whole subject matter of mathematics instruction was the following. In the 3rd class arithmetic and algebra; in the 4th class plane geometry, solid geometry, plane trigonometry, spherical trigonometry, elements of geodesy and cartography; in the 5th class conic sections, solving equations of higher degree, elements of differential and integral calculus, mathematical geography; in the 6th class solid and fluid mechanics (the program of the 7th class, the last one, included no subject of a mathematical nature). All this did not mean anything new to János.

János soon became one of the best students of the academy, and he had time left even for meditating on the problem connected with the fifth axiom. In spring 1820 he wrote to his father that he was making efforts to prove the axiom. The father, in his reply, anxiously tried to divert him from his object and entreated him to deal with less barren questions. Nevertheless, in the same letter he described how much he had struggled in order to clear up the problem himself, and he made a digression on presenting his own ideas. The letter which aimed at dissuasion became fuel to the fire, and János threw himself into the problem with even more ardent interest.

It might have been after receiving the anxious warning from his father that a fruitful idea, which meant the beginning of absolute geometry in János Bolyai's mind, flashed through him. He recalls this first spark in subsequent writings. He writes that as early as in 1820 he stepped on to the way that led him to absolute geometry. He discovered LEGENDRE's angle theorems of himself and he raised the question of what are the consequences of the assumption "the angle sum of the triangle is $<2\varrho$". It first required a new definition of the concept of parallelism. This, in turn, led to a new way of looking at the circle of infinite radius.

From this new aspect, the distance line, that is the path of a point running at a fixed distance from a straight line appeared to be a curve. *This fruitful idea was an element of crucial importance in the formation of János Bolyai's intellectual world.* At the same time, it meant the first phase of his loneliness. In fact, his father could no more follow this idea; he understood the non-Euclidean concept of geometry only after GAUSS' letter of 1832 had convinced him of the correctness and scientific value of the work of his son.

7. THE DISCOVERY OF ABSOLUTE GEOMETRY

First of all, we must point out that although JÁNOS BOLYAI had an excellent education in the basic facts of mathematics, his familiarity with special literature was poor, and he had only a rather incomplete notion of contemporary achievements. Even of GAUSS' results only a small proportion was known to him; for example, he has not heard of the investigations of Gauss in surface theory contained in the work *Disquisitiones generales circa superficies curvas* throughout his life. Neither was he aware of work done by SACCHERI, LAMBERT, SCHWEIKART, and TAURINUS. He got acquainted with the ideas of GAUSS only after the *Appendix* had been published, and he took notice of a (single) work of LOBACHEVSKY even later, in 1848 (these circumstances have already been pointed out by STÄCKEL*).

Obviously, the relative lack of information increased the difficulties encountered by BOLYAI. The stimulating effect of scholarly environment, the interest shown by colleagues working in related fields, acquaintance with the methods of other scientists are all solid supports and strong impulses for an effective and stout research work. Bolyai had no share of them.

The first written document from which we can draw conclusions regarding the progress of his ideas is dated from 1820. In one of BOLYAI's sixth-class copy-books of mechanics there are four illustrations accompanied by the words *Parallelarum Theoria*. They say a lot to the well versed mathematician.

Three of the illustrations are concerned with the (limit) circle of infinite radius, while the fourth shows a "regular octagon" whose sides are straight lines parallel to two principal diagonals each. The metamorphosis of Bolyai's attitude to geometry has already begun.

Finally, one winter night of 1823 he established the relation connecting the so-called angle of parallelism with the distance of parallelism**. This was the second decisive step in his investigations.

Rejoicing over the discovery, a fairly common experience in research work, the complete contour of the structure of non-Euclidean geometry occurred to him. This sudden occurrence of the final form directed his pen when writing the famous letter to this father on 3rd November 1823:

"... *I have created another world, a new world from nothing*: all I had sent before was merely a house of cards as compared with a tower. I am convinced that it will not do much less credit to me than an invention could do."

The father, in his reply, tries to cool the enthusiasm of János. He is anxious for his son since he cannot believe that János has really found a new way which leads him to the solution of the problem open for two thousand years.

In February 1825 János visits his father in Marosvásárhely and shows him the construction of *absolute* space science in detail. The father formally understands, or rather follows and takes note of the ideas of János, but cannot graps their essence and significance. He is unable to get rid of the Euclidean aspect deeply rooted in him. After passionate debates János leaves full of sorrow. He did not succeed in convincing his former intellectual companion.

In 1826 Bolyai hands over a "piece of work" to his former professor, captain JOHANN WOLTER VON ECKWEHR. It "lays the foundations" of absolute geometry

* Cf. P. Stäckel: The Life and Work of the Two Bolyai's

** Cf. *Appendix*, § 29.

(data taken from manuscripts of János Bolyai). Unfortunately, the manuscript sent to ECKWEHR has not been found.

Farkas Bolyai had been writing his principal work for about twenty years when, in 1829, he was granted the permission to publish it in print. It appeared in two volumes, the first one in 1832.

In the autumn of 1830 János met his father again. Probably by arrangement made on this occasion, he decided to write a brief exposition of absolute geometry. He gave the manuscript, in Latin, to his father in 1831 with the purpose of adding it, as one of the appendices, to the father's book just in preparation.

The title of Farkas Bolyai's book was the following: *Tentamen juventutem studiosam in elementa Matheseos ... introducendi.* One of the three appendices to the first volume belonged to János. Its title was: *Appendix scientiam spatii absolute veram exhibens: a veritate aut falsitate Axiomatis XI Euclidei, a priori haud unquam decidenda, indipendentem: adiecta ad casum falsitatis quadratura circuli geometrica.* For brevity, these works are usually referred to as *Tentamen* and *Appendix,* respectively. The offprints of the *Appendix* came out *in June 1831.*

FARKAS BOLYAI sent the *Appendix* to GAUSS and, in the name of his son, asked his opinion. GAUSS' famous reply bore the date 6th March 1832. Farkas had it duplicated and sent a copy to János who was staying in Lemberg. János received the letter on 6th April.

The letter of GAUSS was the first criticism given by an expert. In spite of the cool style, it expressed sincere appreciation. Let us quote that part concerned with the *Appendix* in full.

"Now something about the work of your son. If I begin by saying *that I must not praise him,* surely, you will be startled for a moment; but I cannot do otherwise; praising him would mean praising myself: because all the contents of the work, the way followed by your son, and the results he obtained agree almost from beginning to end with the meditations I had been engaged in partly for 30–35 years already. This extremely surprised me indeed.

It had been my intention to publish nothing of my own work during my life; by the way, I have noted down only a small portion so far. Most people do not even have a right sense of what this matter depends on, and I have met only few to accept with particular interest what I told them. One needs a strong feeling of what in fact is missing and, as to this point, the majority of people lack it. On the other hand, I had planned to write down everything in the course of time so that at least it would not vanish with me some day.

Thus I was greatly surprised that now I can save myself this trouble, and I am very glad that it is just my good old friend's son who so wonderfully outmatched me.

I find the notation highly characteristic and concise; in my opinion, however, it would be good to fix not only symbols or letters, but also definite names for some basic notions, and I have been considering a few such names for a long time. Examining the matter by intuition we need neither names nor symbols; they only become necessary when we want to make ourselves understood by others.

For instance, the surface and the line your son calls F and L might be named parasphere and paracycle, respectively: they are, in essence the sphere and circle of infinite radii. One might call hypercircle the collection of all points at equal distance from a straight line with which they lie in the same plane; similarly for hypersphere. But all these are subordinate questions of no importance; content rather than form is the main point.

In some parts of the investigation I had proceeded in a slightly different way; as a sample, I enclose the (essentially) pure geometrical proof of the theorem which says that the difference from 180° of the angle sum of a triangle is proportional to the area of the triangle.

... I tried here to present just the outlines of the proof without any smoothing or polishing for which I have no time now. You are free to inform your son of it; in any case, please give him my kind regards and assure him of my sincere appreciation. At the same time, invite him to deal with the following problem: *To find the volume of the tetrahedron (space bounded by four planes).*

As the area of the triangle can be obtained so simply, one would expect there is a simple expression also for the volume asked above. This expectation appears to be treacherous.

For treating geometry correctly from the outset, it is indispensable to prove the possibility of the *planum*; the usual definition contains too much and, as a matter of fact, tacitly includes a theorem. It is surprising that all authors from Euclid until recent times have gone about the problem so carelessly. This difficulty, however, is of a quite other nature than that of deciding between Σ and S, and is not very hard to remove. Probably just your book will satisfy me in this respect.

That it is impossible to decide *a priori* between Σ and S is the clearest evidence of the mistake KANT had made when stating that space was merely the *form* of our looking at things. I pointed out another, equally strong, reason in a short paper to be found in the year 1831 volume of the Gött. Gel. Anzeigen as item 64 on p. 625. Perhaps it will not be a disappointment if you try to procure that volume of the G. G. A. (which may be accomplished through any bookseller in Vienna or Buda), as you also find there, developed in a few pages, the essence of my views concerning imaginary quantities."

By a certain claim of priority, GAUSS' letter strongly resembles a statement he had made relative to investigations of JACOBI and ABEL fundamentally important in the theory of elliptic functions. LEGENDRE had attached much value to ABEL's works, and the statement had shocked him. Similarly, JÁNOS BOLYAI had good reason to become indignant at GAUSS' behaviour.

One should agree with Bolyai who has written the following:

"In my opinion and, I firmly believe, in any impartial opinion all of the reasons GAUSS has put forward in order to explain why he did not want to publish anything

3*

of his papers on the subject throughout his life are ineffective and void; really, in science just as in everyday life the question always is to clarify the necessary and useful but still obscure things adequately and to arouse, duly train and promote the turn, missing or dormant, for what is true and correct. To the great loss and detriment of humanity, aptitude for mathematics occurs generally in very few people; for this reason or on this pretext, GAUSS could remain consistent only by concealing another very considerable part of his excellent works at home. Neither the fact that, unfortunately, several mathematicians, even celebrated ones, are superficial, may serve for an intelligent man as a basis to create only superficial and mediocre things and leave science lethargically in the state inherited. Such an assumption may be described as directly unnatural and absolute nonsense; it is all the more to be resented if Gauss, instead of acknowledging the great value of the *Appendix* and the whole *Tentamen* in a straightforward, definite and open manner, expressing his great pleasure and interest, and trying to assure due reception to a good cause, rather avoided all this and confined himself to simple wishes and complaints about the lack of proper culture. Life, activity and merit are certainly different from that."

The following lines written by him in an application dated from 1832 reflect a similar view:

"GAUSS seems to have been the only to make some fairly easy steps towards the goal, but he was still very far from seeing it. In spite of all efforts, however, he was unable to *advance*; the author can make this unquestionable by several data contained partly in the present, partly in the previous letter of GAUSS and several letters of the father (addressed to the author). It was only in the second half of the year 1823 that the author overcame all the main difficulties he had encountered. Formerly, when he had had the honour to be a student of the Imperial and Royal Academy of Engineering, he had been captured by a keen interest in every real knowledge and particularly in this subject which is, apart from its importance and fame, so extraordinary even from the historical point of view. After a few light attempts which had still come short of reaching the final aim, he did not shrink back from the troubles of a powerful attack to that broad and so insatisfactory gap. And he deeply feels that he will not find peace and happiness before winding out of this labyrinth."

GAUSS' attitude towards JÁNOS BOLYAI was unworthy of his own immense scientific prestige deserved by creative work. He was unaware of the moral obligation that to call the attention of scientific community to a great discovery and help the unknown scientist on his way of development is the duty of those who have already reached the summit and whose words of appreciation can give weight to others as well.

*

Subsequently, during the elaboration of absolute geometry, JÁNOS BOLYAI has obtained further important results. The present book deals only with the *Appendix,* BOLYAI's single work published in print. For a long time, no other results of his were known.

He has remained isolated to the end of his life. Not by his fault, but by the social conditions of that time.

PART II

THE ABSOLUTE GEOMETRY OF JÁNOS BOLYAI. THE APPENDIX

APPENDIX.

SCIENTIAM SPATII *absolute verum* exhibens:

a veritate aut falsitate Axiomatis XI *Euclidei* (*a priori haud unquam decidenda*) *independentem*: adjecta ad casum falsitatis, quadratura circuli geometrica.

Auctore JOHANNE BOLYAI de eadem, Geometrarum in Exercitu Caesareo Regio Austriaco Castrensium Capitaneo

---

EXPLICATIO SIGNORUM.

$\widetilde{ab}$ denotet complexum *omnium* punctorum cum punctis a, b in recta sitorum.

$a\tilde{b}$ . . . rectae $\widetilde{ab}$ in a bifariam sectae dimidium illud, quod punctum b complectitur.

$\widetilde{ab}c$ . . . complexum *omnium* punctorum, quae cum punctis a, b, c (non in eadem recta sitis) in eodem plano sunt.

$ab\tilde{c}$ . . . plani $\widetilde{ab}c$ per $\widetilde{ab}$ bifariam secti dimidium, punctum c complectens.

abc . . . portionum, in quas $\widetilde{ab}c$ per complexum rectarum $b\tilde{a}$, $b\tilde{c}$ dividitur, *minorem*; sive *angulum*, cuius $b\tilde{a}$, $b\tilde{c}$ crura sunt.

$abcd$ . . . (si d in abc sit et $b\tilde{a}$, $c\tilde{d}$ se invicem non secent) portionem ipsius abc inter $b\tilde{a}$, $b\tilde{c}$, $c\tilde{d}$ comprehensam; $bacd$ vero portionem plani $\widetilde{ab}c$ inter $\widetilde{ab}$, $c\tilde{d}$ sitam.

$\perp$ . . . perpendiculare.

$\wedge$ . . . angulum.

R . . . angulum rectum.

$ab \triangleq cd$. $cab = acd$.

$\equiv$ . . . congruens *).

$x \frown a$. x tendere ad limitem a.

$\bigcirc r$ . . peripheriam circuli radii r.

$\odot r$ . . aream circuli radii r.

*) Sit fas, signo hocce, quo summus Geometra GAVSS *numeros congruos* insignivit; congruentiam geometricam quoque denotare: nulla' ambiguitate exinde metuenda.

§ 1. (Fig. 1.) Si rectam $\widetilde{am}$ non secet plani ejusdem recta $\widetilde{bn}$, at secet quaevis $\widetilde{bp}$ (in abn): designetur hoc per $\widetilde{bn} \,|||\, \widetilde{am}$. *Dari* talem $\widetilde{bn}$, et quidem *unicam*, e quovis puncto b (extra $\widetilde{am}$), atque $bam \dagger abn$ non $> 2R$ esse patet; nam bc circa b mota, donec $bam \dagger abc = 2R$ fiat, $\widetilde{bc}$ ex $\widetilde{am}$ aliquando *primum* exit, estque tunc $\widetilde{bc} \,|||\, \widetilde{am}$. Nec non patet esse $\widetilde{bn} \,|||\, \widetilde{em}$, ubivis sit e in $\widetilde{am}$ (supponendo in omnibus talibus casibus esse $am > ae$). Et si, puncto c in $\widetilde{am}$ abeunte in infinitum, semper sit $cd = cb$: erit semper $cdb = (cbd < nbc)$; ast $nbc \frown o$; adeoque et $adb \frown o$.

§ 2. (Fig. 2). Si $\widetilde{bn} \,|||\, \widetilde{am}$; est quoque $\widetilde{cn} \,|||\, \widetilde{am}$. Nam sit d ubicunque in $macn$. Si c in $\widetilde{bn}$ sit; $\widetilde{bd}$ secat $\widetilde{am}$ (propter $\widetilde{bn} \,|||\, \widetilde{am}$), adeoque et $\widetilde{cd}$ secat $\widetilde{am}$; si vero c in $\widetilde{bp}$ fuerit; sit $\widetilde{bq} \,|||\, \widetilde{cd}$: cadit $\widetilde{bq}$ in abn (§ 1), secatque $\widetilde{am}$, adeoque et $\widetilde{cd}$ secat $\widetilde{am}$. Quaevis $\widetilde{cd}$ igitur (in acn) secat in utroque casu $\widetilde{am}$ absque eo, ut $\widetilde{cn}$ ipsam $\widetilde{am}$ secet. Est ergo semper $\widetilde{cn} \,|||\, \widetilde{am}$.

§ 3. (Fig. 2). Si tam $\widetilde{br}$ quam $\widetilde{cs}$ sit $|||\, \widetilde{am}$, et c non sit in $\widetilde{br}$; tum $\widetilde{br}, \widetilde{cs}$ se invicem haud secant. Si enim $\widetilde{br}, \widetilde{cs}$ punctum d commune haberent; (per §. 2.) essent $\widetilde{dr}$ et $\widetilde{ds}$ simul $|||\, \widetilde{am}$, caderetque (§. 1.) $\widetilde{ds}$ in $\widetilde{dr}$ et c in $\widetilde{br}$ (contra hyp).

§ 4. (Fig. 3) Si $man > mab$; pro quovis puncto b ipsius $\widetilde{ab}$ *datur* tale c in $\widetilde{am}$, ut sit $bcm = nam$. Nam datur per (§. 1.) $bdm > nam$, adeoque $mdp = man$, caditque b in $nadp$. Si igitur nam juxta am feratur, usquequo $\widetilde{an}$ in $\widetilde{dp}$ veniat; aliquando $\widetilde{an}$ per b transiisse, et aliquod $bcm = ncm$ esse oportet.

§ 5. (Fig. 1). Si $bn|||am$, *datur* tale punctum f in $\widetilde{am}$, ut sit $fm \simeq bn$. Nam per §. 1. datur $bcm >$ cbn; et si $ce = cb$, adeoque $ec \simeq bc$; patet esse $bem < ebn$. Feratur p per ec, $\wedge$lo bpm semper u, et $\wedge$lo pbn semper v dicto; patet u esse prius ei simultaneo v minus, posterius vero esse majus. Crescit vero u a bem usque bcm *continuo*; cum (per §. 4.) *nullus* $\wedge$lus $> bem$ et $< bcm$ detur, cui u aliquando $=$ non fiat; pariter decrescit v ab ebn usque cbn continuo: datur itaque in ec tale f, ut $bfm = fbn$ sit.

§ 6. Si $bn|||am$, atque ubivis sit e in $\widetilde{am}$, et g in $\widetilde{bn}$: tum $gn|||em$ et $em|||gn$. Nam (per §. 1.) est $bn|||em$, et hinc (per §.2.) $gn|||em$. Si porro $fm \simeq bn$ (§.5.); tum $mfbn \equiv nbfm$, adeoque (cum $bn|||fm$ sit) etiam $fm|||bn$, et (per praec.) $em|||gn$.

§ 7. (Fig. 4). Si tam bn quam cp sit $|||am$, et c non sit in $\widetilde{bn}$: est etiam $bn|||cp$. Nam $\widetilde{bn}$, $\widetilde{cp}$ se invicem non secant (§. 3); sunt vero am, bn, cp aut in plano, aut non; atque in casu primo am aut in $bncp$ est, aut non. Si am, bn, cp in plano sint, et am in $bncp$ cadat; tum quaevis $\widetilde{bq}$ (in nbc) secat $\widetilde{am}$ in aliquo puncto d (quia $bn|||am$); porro cum $dm|||cp$ sit (§. 6.), patet $\widetilde{dq}$ secare cp, adeoque esse $bn|||cp$. Si vero bn, cp in eadem plaga ipsius am sint; tum aliqua earum ex. gr. cp, *intra* duas reliquas $\widetilde{bn}$, $\widetilde{am}$ cadit; quaevis $\widetilde{bq}$ (in nba) autem secat $\widetilde{am}$, adeoque et ipsam $\widetilde{cp}$. Est itaque $bn|||cp$.

Si mab, mac, $\wedge$lum efficiant; tum cbn cum abn nonnisi $\widetilde{bn}$, $\widetilde{am}$ vero (in abn) cum $\widetilde{bn}$, adeoque nbc quoque cum $\widetilde{am}$, nihil commune habent. Per quamvis $\widetilde{bd}$ (in nba) autem positum bcd secat $\widetilde{am}$, quia (propter $bn|||am$) $\widetilde{bd}$ secat $\widetilde{am}$. Moto itaque bcd circn *bc*, donec ipsam $\widetilde{am}$ *prima vice* deserat, postremo cadet $\widetilde{bcd}$ in $\widetilde{bcn}$. Eadem ratione cadet idem in $\widetilde{bcp}$; cadit igitur *bn* in *bcp*. Porro si *br* ||| *cp*; tum (quia etiam *am* ||| *cp*) pari ratione cadit *br* in *bam*; nec non (propter *br* ||| *cp*) in *bcp*. Itaque $\widetilde{br}$ ipsis *mab*, *pcb* commune, nempe ipsum $\widetilde{bn}$ est, atque hinc *bn* ||| *cp*.

Si igitur *cp* ||| *am*, et *b* extra $\widetilde{cam}$ sit: tum sectio ipsorum *bam*, *bcp*, nempe $\widetilde{bn}$, est ||| tam ad *am*, quam ad *cp*.

§ 8. (Fig.5). Si *bn* ||| et ≏ *cp* (vel brevius *bn* ||| ≏ *cp*), atque *am* (in *nbcp*) rectam *bc* ⊥riter bissecet; tum *bn* ||| *am*. Si enim $\widetilde{bn}$ secaret $\widetilde{am}$, etiam $\widetilde{cp}$ secaret $\widetilde{am}$ in eodem puncto (cum *mabn* ≡ *macp*), quod et ipsis $\widetilde{bn}$, $\widetilde{cp}$ commune esset, quamvis *bn* ||| *cp* sit. Quaevis $\widetilde{bq}$ (in *cbn*) vero secat $\widetilde{cp}$; adeoque secat $\widetilde{bq}$ etiam $\widetilde{am}$. Consequenter *bn* ||| *am*.

§ 9. (Fig 6). Si *bn* ||| *am*, *map* ⊥ *mab*, atque ∧, quem *nbd* cum *nba* (in ea plaga ipsius *mabn*, ubi *map* est) facit, sit < *R*: tum *map* et *nbd* se invicem secant. Nam sit *bam* = *R*, *ac* ⊥ $\widetilde{bn}$ (sive in *b* cadat *c*, sive non) et *ce* ⊥ *bn* (in *nbd*); erit (per hyp.) *ace* < *R*, et *af* (⊥ *ce*) in *ace* cadet. Sit $\widetilde{ap}$ sectio (punctum *a* commune habentium) $\widetilde{abf}$ et $\widetilde{amp}$; erit *bap* = *bam* = *R* (cum sit *bam* ⊥ *map*). Si denique $\widetilde{abf}$ in $\widetilde{abm}$ ponatur (*a* et *b* manentibus); cadet $\widetilde{ap}$ in $\widetilde{am}$; atque cum *ac* ⊥ *bn* et *af* ⊥ *ac* sit, patet *af intra* $\widetilde{bn}$ terminari, adeoque *bf* in *abn* cadere. Secat autem $\widetilde{bf}$ ipsam $\widetilde{ap}$ in *hoc* situ (quia *bn* ||| *am*), adeoque etiam in situ *primo*, $\widetilde{ap}$ et $\widetilde{bt}$ se invicem secant; estque punctum sectionis ipsis $\widetilde{map}$ et $\widetilde{nbd}$ commune: secant itaque $\widetilde{map}$ et $\widetilde{nbd}$ se invicem. Facile exhinc sequitur $\widetilde{map}$ et $\widetilde{nbd}$ se mutuo secare, si summa internorum, quos cum *mabn* efficiunt, $< 2R$ sit.

§ 10. (Fig.7). Si tam bn quam cp sit $|||\bumpeq am$: est etiam $bn\,|||\bumpeq cp$. Nam mab et mac aut $\wedge$*lum*. efficiunt, aut in plano sunt.

Si prius; bissecet $\widetilde{qdf}$ rectam ab $\perp$*riter*; erit $dq \perp ab$, adeoque $dq\,|||\,am$ (§.8.); pariter si $\widetilde{ers}$ bissecet rectam ac $\perp$*riter*, est $er\,|||\,am$; unde $dq\,|||\,er$ (§. 7.). Facile hinc (per §. 9.) consequitur, $\widetilde{qdf}$ et $\widetilde{ers}$ se mutuo secare, et sectionem $\widetilde{fs}$ esse $|||\,dq$ (§. 7.), atque (propter $bn\,|||\,dq$) esse etiam $fs\,|||\,bn$. Est porro (pro quovis puncto ipsius $\widetilde{fs}$) $fb = fa = fc$, caditque $\widetilde{fs}$ in planum $\widetilde{tgf}$, rectam bc $\perp$*riter* bissecans. Est vero (per §. 7.) (cum sit $fs\,|||\,bn$) etiam $gt\,|||\,bn$. Pari modo demonstratur $gt\,|||\,cp$ esse. Interim gt bissecat rectam bc $\perp$*riter*; adeoque $tbgn \equiv tgcp$ (§.1.) et $bn\,|||\bumpeq cp$.

Si bn, am, cp in plano sint; sit (*extra* hoc planum cadens) $fs\,|||\bumpeq am$; tum (per praec.) $fs\,|||\bumpeq$ tam ad bn quam ad cp, adeoque et $bn\,|||\bumpeq cp$.

§ 11. Complexus puncti a, atquo *omnium* punctorum, quorum quodvis b tale est, ut si $bn\,|||\,am$ sit, sit etiam $bn \bumpeq am$; dicatur F: sectio vero ipsius E cum quovis plano rectam am complectente nominetur L. In quavis recta, quae $|||\,am$ est, F gaudet puncto, et nonnisi uno; atque patet L per $\widetilde{am}$ dividi in duas partes congruentes; dicatur $\widetilde{am}$ *axis* ipsius L; patet etiam, in quovis plano rectam am complectente, pro *axe* $\widetilde{am}$ unicum L dari. Quodvis eiusmodi L, dicatur L *ipsius* $\widetilde{am}$ (in plano, de quo agitur, intelligendo). Patet per L circa am revolutum, F describi, cuius $\widetilde{am}$ axis vocetur, et vicissim F *axi* $\widetilde{am}$ *attribuatur*.

§ 12. Si b ubivis in L ipsius $\widetilde{am}$ fuerit, et $\widetilde{bn} \parallel\!| \triangleq \widetilde{am}$ (§. 11); tum L ipsius $\widetilde{am}$ et L ipsius $\widetilde{bn}$ *coincidunt*. Nam dicatur L ipsius $\widetilde{bn}$ distinctionis ergo l; sitque c ubivis in l, et $cp \parallel\!| \triangleq bn$ (§. 11.); erit (cum et $bn \parallel\!| \triangleq am$ sit) $cp \parallel\!| \triangleq am$ (§. 10), adeoque c etiam in L cadet. Et si c ubivis in L sit, et $cp \parallel\!| \triangleq am$; tum $cp \parallel\!| \triangleq bn$ (§. 10.); caditque c etiam in l (§. 11). Itaque L et l sunt eadem; ac quaevis $\widetilde{bn}$ est etiam axis ipsius L, et inter omnes axes ipsius L, $\triangleq$ est. Idem de F eodem modo patet.

§. 13. (Fig. 8). Si $bn \parallel\!| am$, $cp \parallel\!| dq$, et $bam + abn = 2R$ sit; tum etiam $dcp + cdq = 2R$. Sit enim $ea = eb$ et $efm = dcp$ (§. 4.); erit (cum $bam + abn = 2R = abn + abg$ sit) $ebg = eaf$; adeoque si etiam $bg = af$ sit, $\triangle ebg \equiv \triangle eaf$, $beg = aef$, cadetque g in $\widetilde{fe}$. Est porro $gfm + fgn = 2R$ (quia $egb = efa$). Est etiam $gn \parallel\!| fm$ (§. 6.); itaque si $mfrs \equiv pcdq$, tum $rs \parallel\!| gn$ (§. 7.), et r in vel extra fg cadit (si cd non $= fg$, ubi res jam patet).

I. In casu primo est frs non $>$ $(2R - rfm = fgn)$, quia $rs \parallel\!| fm$; ast cum $rs \parallel\!| gn$ sit, est etiam frs non $<$ fgn; adeoque $frs = fgn$, et $rfm + frs = gfm + fgn = 2R$. Itaque et $dcp + cdq = 2R$.

II. Si r extra fg cadat; tunc $ngr = mfr$, sitque $mfgn \equiv nghl \equiv lhko$ et ita porro, usquequo fk = vel prima vice $> fr$ fiat. Est heic $ko \parallel\!| hl \parallel\!| fm$ (§. 7.). Si k in r cadat; tum ko in rs cadit (§. 1.); adeoque $rfm + frs = kfm + fko = kfm + fgn = 2R$; si vero r in hk cadat, tum (per I.) est $rhl + krs = 2R = rfm + frs = dcp + cdq$.

§ 14. Si $bn \parallel\!| am$, $cp \parallel\!| dq$, et $bam + abn < 2R$ sit; tum etiam $dcp + cdq < 2R$. Si enim $dcp + cdq$ non esset $<$, adeoque (per §. 1.) esset $= 2R$; tum (per §. 13.) etiam $bam + abn = 2R$ esset (contra hyp).

§ 15. Perpensis §§. 13. et 14. *Systema Geometriae, hypothesi veritatis Axiomatis Euclidei* XI *insistens dicatur* Σ; *et hypothesi contrariae super-*

structum sit S. *Omnia, quae expresse non dicentur, in* Σ vel *in* S *esse; absolute enuntiari*, i. e. *illa, sive* Σ *sive* S *reipsa sit, vera asseri intelligatur.*

§ 16. (Fig. 5). Si am sit axis alicujus L; tum L in Σ *recta* $\perp am$ est. Nam sit e quovis puncto b ipsius L axis bn; erit in Σ $bam + abn = 2bam = 2R$, adeoque $bam = R$. Et si c quodvis punctum in $\widetilde{ab}$ sit, atque $cp \,|||\, am$; est (per §. 13.) $cp \triangleq am$, adeoque c in L (§. 11.)

In S vero *nulla* 3 puncta a, b, c ipsius L vel F in recta sunt. Nam aliquis axium am, bn, cp (ex.gr. am) intra duos reliquos cadit; et tunc (per §. 14.) tam bam quam $cam < R$.

§ 17. *L est etiam in S linea, et F superficies.* Nam (per §. 11.) quodvis planum ad axem am (per punctum aliquod ipsius F) $\perp$re, secat ipsum F in peripheria circuli, cuius planum (per §. 14.) ad nullum alium axem $\widetilde{bn}$ $\perp$re est. Revolvatur F circa bn; manebit (per §. 12.) quodvis punctum ipsius F in F, et sectio ipsius F cum plano ad $\widetilde{bn}$ non $\perp$ri, describet superficiem: atqui F (per §. 12), quaecunque puncta a, b fuerint in eo, ita *sibi* congruere poterit, ut a in b cadat; est igitur F *superficies uniformis*. Patet hinc (per §. 11. et 12) L esse *lineam uniformem*.

§ 18. (Fig. 7). *Cujusvis plani*, per punctum a ipsius F ad axem am oblique positi, *sectio* cum F in S *peripheria circuli* est. Nam sint a, b, c, 3 puncta hujus sectionis, et bn, cp axes; facient $ambn$, $amcp$ $\wedge$*lum*; nam secus planum (ex §. 16.) per a, b, c determinatum ipsam am complecteretur (contra hyp). Plana igitur, rectas ab, ac $\perp$*riter* bissecantia se mutuo secant (§. 10.) in aliquo *axe* $\widetilde{fs}$ (ipsius F), atque $fb = fa = fc$. Sit $ah \perp fs$, et revolvatur fah circa fs; describet a peripheriam radii ha, per b et c euntem, et *simul* in F et $\widetilde{abc}$ sitam nec F et $\widetilde{abc}$ praeter ○ *ha* quidquam commune habent (§. 16.). Patet etiam portione *fa* lineae L (tanquam radio) in F circa *f* mota ipsam ○ *hh* describi.

§ 19. (Fig.5). $\perp$ris *bt* ad axem *bn* ipsius L (in planum ipsius L cadens) est in S *tangens* ipsius L. Nam L in $\widetilde{bt}$ praeter *b* nullo puncto gaudet (§.14.), si vero *bq* in *tbn* cadat, tum centrum sectionis plani per *bq* ad *tbn* $\perp$ris cum F ipsius $\widetilde{bn}$ (§.18.) manifesto in $\widetilde{bq}$ locatur, et si *bc* diameter sit, patet $\widetilde{bq}$ lineam L ipsius $\widetilde{bn}$ in *c* secare.

§. 20. Per quaevis 2 puncta in F linea *L* determinatur (§. 11. et 18); atque (cum ex §§. 16. et 19 . L $\perp$ ad omnes suos axes sit) quivis $\angle$ L lineis in F, $\angle$lo planorum ad F per *crura* $\perp$rium, = est.

§ 21. (Fig. 6). Duae lineae Lformes $\widetilde{ap}$, $\widetilde{bd}$ in, eodem F, cum tertia Lformi *ab* summam internorum < 2R efficientes, se mutuo secant (per $\widetilde{ap}$ in F intelligendo *L* per *a*,*p* ductum, per $\widetilde{ap}$ vero dimidium illud eius ex *a* incipiens, in quod *p* cadit). Nam si *am*, *bn* axes ipsius F sint; tum $\widetilde{amp}$. $\widetilde{bnd}$ secant se invicem (§.9.); atque F secat eorundem sectionem (per §§.7. et 11.); adeoque et $\widetilde{ap}$, $\widetilde{bd}$ se mutuo secant.

Patet exhinc Axioma XI. et omnia, quae in Geometria Trigonometriaque (plana) asseruntur, *absolute* constare in *F*, rectarum vices lineis *L* subeuntibus: idcirco functiones trigonometricae abhinc eodem sensu accipientur, quo in Σ veniunt; et peripheria circuli, cuius radius Lformis $= r$ in *F*, est $= 2\pi r$, et pariter ⊙ r (in *F*) $= \pi r^2$ (per π intelligendo $\frac{1}{2}$ ○1 in *F*, sive notum 3,1415926...)

§ 22. (Fig.9. Si $\widetilde{ab}$ fuerit *L* ipsius $\widetilde{am}$, et *c* in $\widetilde{am}$; atque $\angle$ *cab* (e recta $\widetilde{am}$ et Lformi linea

$\widetilde{ab}$ compositus) feratur prius juxta $\widetilde{ab}$, tum juxta $\widetilde{ba}$ semper porro in infinitum: erit via $\widetilde{cd}$ ipsius c linea L ipsius cm. Nam (posteriori l dicta) sit punctum quodvis d in $\widetilde{cd}$, $dn \,|||\, cm$, et b punctum ipsius L in $\widetilde{dn}$ cadens: erit $bn \triangleq am$. et $ac = bd$, adeoque $dn \triangleq cm$, consequ. d in l. Si vero d in l et $dn \,|||\, cm$, atque b punctum ipsius L ipsi $\widetilde{dn}$ commune sit: erit $am \triangleq bn$ et $cm \triangleq dn$, unde manifesto $bd = ac$, cadetque d in viam puncti c, et sunt l et $\widetilde{cd}$ eadem. Designetur tale l per $l \parallel$ L.

§. 23. (Fig.9) Si linea Lformis $cdf \parallel abe$ (§.22.), et $ab = be$, atque $\widetilde{am}, \widetilde{bn}, \widetilde{ep}$ sint axes: erit manifesto $cd = df$: et si quaelibet 3 puncta a, b, e fuerint ipsius $\widetilde{ab}$, ac $ab = n.cd$. erit quoque $ae = n.ef$; adeoque (manifesto etiam pro ab, ae, dc incommensurabilibus) $ab : cd = ae : ef$, estque $ab : cd$ ab ab *independens*, et per ac *prorsus determinatum*. Denotetur quotus iste, nempe $ab : cd$ litera majori eiusdem nominis (puta per X), quo ac litera minuscula (ex.gr. x) insignitur.

24. Quaecunque x et y fuerint; est $X = Y^{\frac{y}{x}}$ (§.23.) Nam aut erit alterum (ipsorum x, y) multiplum alterius (ex.gr. y ipsius x), aut non.

Si $y = nx$; sit $x = ac = cg = gh$ &c, usque quo $ah = y$ fiat; sit porro $cd \parallel gk \parallel hl$; erit (§.23.) $X = ab : cd = cd : gk = gk : hl$; adeoque $\frac{ab}{hl} = \left(\frac{ab}{cd}\right)^n$, sive $Y = X^n = X^{\frac{y}{x}}$. Si x, y multipla ipsius i sint, puta $x = mi$, et $y = ni$; est (per praec.) $X = I^m$, $Y = I^n$, consequ. $Y = X^{\frac{n}{m}} = X^{\frac{y}{x}}$. Idem ad casum incommensurabilitatis ipsorum x, y facile extenditur. Si vero fuerit $q = y - x$; erit manifesto $Q = Y : X$.

Nec non manifestum est, in Σ pro quovis x esse $X=1$, in S vero $X>1$ esse, atque pro *quibusvis* ab, abe dari tale $cdf \parallel abe$, ut sit $cdf=ab$, unde $ambn \equiv amep$ erit, etsi hoc illius qualevis multiplum sit; quod singulare quidem est, sed absurditatem ipsius S evidenter non probat.

§.25. (Fig.10) *In quovis rectilineo △lo sunt peripheriae radiorum lateribus aequalium, uti sinus ∧lorum oppositorum.*

Sit enim $abc=R$, et $am \perp bac$, atque sint bn, $cp \parallel\parallel am$; erit $cab \perp ambn$, adeoque (cum $cb \perp ba$ sit) $cb \perp ambn$, consequ. $cpbn \perp ambn$. Secet F *ipsius* $\widetilde{cp}$, rectas $\widetilde{bn}$, $\widetilde{am}$ (respective) in d, e, et fascias $cpbn$, $cpam$, $bnam$ in lineis Lformibus cd, ce, de; erit (§. 20.) $cde=$ ∧lo ipsorum ndc, nde, adeoque $=R$; atque pari ratione est $ced=cab$. Est autem (per §.21.) in Lineo △ ced (heic radio semper $=1$ posito) $ec:dc=1:\sin dec=1:\sin cab$. Est quoque (per §.21.) $ec:dc=○ec:○dc$ (in F) $=○ac:○bc$ (§. 18.); adeoque est etiam $○ac:○bc=1:\sin cab$; unde assertum pro quovis △lo liquet.

§. 26. *In quovis sphaerico △lo sunt sinus laterum, uti sinus ∧lorum iisdem oppositorum.*

Fig. 11. Nam sit $abc=R$, et $ced \perp$ ad sphaerae radium oa; erit $ced \perp aob$, et (cum etiam $boc \perp boa$ sit) $cd \perp ob$. In △△ ceo, cdo vero est (per §. 25.) $○ec:○oc:○dc=\sin coe:1:\sin cod=\sin ac:1:\sin bc$; interim (§.25.) etiam $○ec:○dc=\sin cde:\sin ced$; Itaque $\sin ac:\sin bc=\sin cde:\sin ced$; est vero $cde=R=cba$, atque $ced=cab$. Consequenter $\sin ac:\sin bc=1:\sin a$. *E quo promanans Trigonometria sphaerica, ab Axiomate* XI *independenter stabilita est*

§. 27. (Fig. 12.) Si ac, bd sint $\perp ab$, et feratur cab juxta $\widetilde{ab}$; erit (via puncti c dicta heic cd) $:ab=\sin u:\sin v$. Nam sit $de \perp ca$; est in △△ ade, adb (per§25.) $○ed:○ad:○ab=\sin u:1:\sin v$. Revoluto $bacd$ circa ac, describetur $○ab$ per b, $○d$ per d; et via dictae cd denotetur heic per $⊙ds$.

Sit porro polygonum quodvis $bfg \cdots$ ipsi $\bigcirc ab$ inscriptum: nascetur per plana ex omnibus lateribus $bf . fg$ &, ad $\odot ab$ $\llcorner$ria, in $\odot cd$ quoque figura polygonalis totidem laterum; et demonstrari ad instar §. 23 potest, esse $cd : ab = dh : bf = hk \cdot fg$ &, adeoque $dh + hk$ & $: bf + fg$ & $= cd : ab$. Quovis laterum $bf, fg \cdots$ ad limitem o tendente, manifesto $bf + fg \cdots \frown \bigcirc ab$, et $dh + hk \cdots \frown \bigcirc cd$. Itaque etiam $\bigcirc cd : \bigcirc ab = cd : ab$. Erat vero $\bigcirc cd : \bigcirc ab = \sin v : \sin v$. Conseq. $cd : ab = \sin u : \sin v$.

Remoto ac a bd. in infinitum, manet $cd : ab$, adeoque etiam $\sin u : \sin v$ *constans*; u vero $\frown R$ (§.1.), et si $dm \,|||\, bn$ sit, $v \frown z$; unde fit $cd : ab = 1 : \sin z$. Via dicta cd denotabitur per $cd \,||\, ab$.

§. 28. (Fig. 13.) Si $bn \,|||\!\simeq \tilde{am}$, et c in $\tilde{am}$, atque $ac = x$ sit: erit X (§.23.) $= \sin u : \sin v$. Nam si cd et ae sint $\llcorner\ bn$, et $bf \llcorner\ am$; erit (ad instar §. 27.) $\bigcirc bf : \bigcirc cd = \sin u : \sin v$. Est autem evidenter $bf = ae$; quamobrem $\bigcirc ea : \bigcirc lc = \sin u : \sin v$. In superficiebus vero Fformibus ipsorum am et cm (ipsum $ambn$ in ab et cg secantibus) est (per §.21.) $\bigcirc ea : \bigcirc dc = ab : cg = X$. Est itaque etiam $X = \sin u : \sin v$.

§.29. (Fig.14.) Si $bam = R$, $ab = y$, et $bn \,|||\, am$ sit; erit in S, $Y = \cot \frac{1}{2} u$. Nam si fuerit $ab = ac$, et $cp \,|||\, am$ (adeoque $bn \,|||\!\simeq cp$), atque $pcd = qcd$; datur (§.19.) $ds \llcorner\ \tilde{cd}$, ut $ds \,|||\, cp$, adeoque (§. 1.) $dt \,|||\, cq$ sit. Si porro $be \llcorner \tilde{ds}$; erit (§. 7.) $ds \,|||\, bn$, adeoque (§.6.) $bn \,|||\, es$, et (cum $dt \,|||\, cg$ sit) $bq \,|||\, et$; consequ. (§. 1.) $ebn = ebq$. Repraesententur, bcf ex L ipsius bn, et fg, dh, ck et el. ex Lformibus lineis ipsorum ft, dt, cq et et; erit evidenter (§.22.) $hg = df = dk = hc$; itaque $cg = 2ch = 2v$. Pariter patet, $bg = 2bl = 2z$ esse. Est vero $bc = bg - cg$; quapropter $y = z - v$, adeoque (§.24.) $Y = Z : V$. Est demum (§. 28.) $Z = 1 : \sin \frac{1}{2} v$, et $V = 1 : \sin (R - \frac{1}{2} v)$ consequ. $Y = \cot . \frac{1}{2} u$.

§. 30. (Fig. 15.) Verumtamen facile (ex § 25) patet, resolutionem problematis *Trigonometriae planae* in S, peripheriae per radium expressae indigere; hoc vero rectificatione ipsius L obtineri potest. Sint ab, cm, $c'm'$ $\perp$ $\widetilde{ac}$, atque b ubivis in $\widetilde{ab}$; erit (§.25.) $\sin u : \sin v := \odot p : \odot y$, et $\sin u' : \sin v' = \odot p : \odot y'$; adeoque $\frac{\sin u}{\sin v} \cdot \odot y = \frac{\sin u'}{\cos v'} \cdot \odot y'$. Est vero (per § 27) $\sin v : \sin v' = \cos u : \cos u'$; conseq. $\frac{\sin u}{\cos u} \odot y = \frac{\sin u'}{\cos u'} \odot y'$; seu $\odot y : \odot y' = \text{tang}\, u' : \text{tang}\, u = \text{tang}\, w : \text{tang}\, w'$ Sint porro cn, $c'n' \parallel\!| \, ab$, et cd, $c'd'$ lineae Lformes ad $\widetilde{ab}$ $\perp$res; erit (§.21.) etiam $\odot y : \odot y' = r : r'$, adeoque $r : r' = \text{tang}\, w : \text{tang}\, w'$. Crescat iam p ab a incipiendo in infinitum; tum $w \frown z$, et $w' \frown z'$; quapropter etiam $r : r' = \text{tang}\, z : \text{tang}\, z'$. *Constans* $r : \text{tang}\, z$ (ab r *independens*) dicatur i; dum $y \frown o$, est $\left(\frac{r}{y} = \frac{i\, \text{tang}\, z}{y}\right) \frown 1$, adeoque $\frac{y}{\text{tang}\, z} \frown i$. Ex §. 29 fit $\text{tang}\, z = \frac{1}{2}\,(Y - Y^{-1})$; itaque $\frac{2y}{Y - Y^{-1}} \frown i$,

seu (§.24.) $\dfrac{2y\, I^{\frac{y}{i}}}{I^{\frac{2y}{i}} - 1} \frown i$.

Notum autem est, expressionis istius (dum $y \frown o$) limitem esse $\frac{\ }{\text{log nat}\, I}$; est ergo $\frac{i}{\text{log nat}\, I} = i$, et $I = e = 2,7182818\cdots$, quae quantitas insignis hic quoque elucet. Si nempe abhinc i illam rectam denotet, cuius $I = e$ sit, erit $r = i\, \text{tang}\, z$. Erat autem (§.21.) $\odot y = 2\pi r$; est igitur $\odot y = 2\pi i$

$$\text{tang}\, z = \pi i (Y - Y^{-1}) = \pi i \left(e^{\frac{y}{i}} - e^{\frac{-y}{i}}\right) = \frac{\pi y}{\text{log nat}\, Y} (Y - Y^{-1})$$ (per §.24.)

§. 31. (Fig. 16.) Ad resolutionem omnium $\triangle$*lorum* rectangulorum rectilineorum trigonometricam (e qua omnium $\triangle$*lorum* resolutio in promtu est) in S,

3 aequationes sufficiunt: nempe (a, b cathetos, c hypotenusam, et α, β $\wedge$*los* cathetis oppositos denotantibus) aequatio relationem exprimens 1*mo* inter a, b, α; 2*do* inter a, α, β; 3*tio* inter a, b, c; nimirum ex his *reliquae* 3 per eliminationem prodeunt.

I. Ex §.25. et 30. est $1 \cdot \sin \alpha = (C - C^{-1}) : (A - A^{-1}) = (e^{\frac{c}{i}} - e^{\frac{-c}{i}}) : (e^{\frac{a}{i}} - e^{\frac{-a}{i}})$ (aequatio pro α; c, a).

II. Ex §.27. sequitur (si $\beta m ||| \gamma n$ sit) $\cos \alpha : \sin \beta = 1 : \sin u$; ex §. 29 autem fit $1 : \sin u = \frac{1}{2}(A + A^{-1})$; itaque $\cos \alpha : \sin \beta = \frac{1}{2}(A + A^{-1}) = \frac{1}{2}(e^{\frac{a}{i}} + e^{\frac{-a}{i}})$ (aequatio pro α, β, a).

III. Si $\alpha\alpha' \perp \beta\alpha\gamma$, atque $\beta\beta'$ et $\gamma\gamma'$ fuerint $\parallel \alpha\alpha'$, (§.27), atque $\beta'\alpha'\gamma' \perp \alpha\alpha'$; erit manifesto (uti in (§. 27) $\frac{\beta\beta'}{\gamma\gamma'} = \frac{1}{\sin u} = \frac{1}{2}(A + A^{-1})$; $\frac{\gamma\gamma'}{\alpha\alpha'} = \frac{1}{2}(B + B^{-1})$, ac $\frac{\beta\beta'}{\alpha\alpha'} = \frac{1}{2}(C + C^{-1})$; consequ. $\frac{1}{2}(C + C^{-1}) = \frac{1}{2}(A + A^{-1}) \cdot \frac{1}{2}(B + B^{-1})$, sive $(e^{\frac{c}{i}} + e^{\frac{-c}{i}}) = \frac{1}{2}(e^{\frac{a}{i}} + e^{\frac{-a}{i}})(e^{\frac{b}{i}} + e^{\frac{-b}{i}})$ (aequatio pro a, b, c).

§.32. Si $\gamma\alpha\delta = R$, et $\beta\delta \perp \alpha\delta$ sit; erit $\bigcirc c : \bigcirc a = 1 : \sin \alpha$, et $\bigcirc c : \bigcirc(d = \beta\delta) = 1 : \cos \alpha$, adeoque ($\bigcirc x^2$ pro quovis x factum $\bigcirc x . \bigcirc x$ denotante) manifesto $\bigcirc a^2 + \bigcirc d^2 = \bigcirc c^2$. Est vero (per §.27. et II.) $\bigcirc d = \bigcirc b . \frac{1}{2}(A + A^{-1})$, consequ. $(e^{\frac{c}{i}} - e^{\frac{-c}{i}})^2 = \frac{1}{4}\left[e^{\frac{a}{i}} + e^{\frac{-a}{i}}\right]^2 . \left[e^{\frac{b}{i}} - e^{\frac{-b}{i}}\right]^2 + \left[e^{\frac{a}{i}} - e^{\frac{-a}{i}}\right]^2$, *alia* aequatio pro a, b, c, (cuius membrum 2*dum* facile ad formam *symmetricam* seu *invariabilem* reducitur.) Denique ex $\frac{\cos\alpha}{\sin\beta}=\frac{1}{2}(A+A^{-1})$, atque $\frac{\cos\beta}{\sin\alpha}=\frac{1}{2}(B+B^{-1})$, fit (per III.) $\cot\alpha.\cot\beta$ $=\frac{1}{2}(e^{\frac{c}{i}}+e^{\frac{-c}{i}})$ (aequatio pro α, β, c).

§. 32. Restat adhuc modum *problemata* in S resolvendi breviter ostendere, quo (per exempla magis obvia) peracto, demum quid theoria haecce praestet, candide dicetur.

I. (Fig.17.) Sit $\widetilde{ab}$ linea in plano, et $y=f(x)$ aequatio eius (pro coordinatis ⊥ *ribus*), et quodvis incrementum ipsius z dicatur dz, atque incrementa ipsorum x, y, et areae u, eidem dz respondentia, respective per dx, dy, du denotentur; sitque bh ∥ cf, et exprimatur (ex §. 31.) $\frac{bh}{dx}$ per y, ac quaeratur ipsius $\frac{dy}{dx}$ *limes* tendente dx ad limitem o, (quod ubi eiusmodi limes quaeritur, subintelligatur): innotescet exinde etiam limes ipsius $\frac{dy}{bh}$, adeoque tang hba; eritque (cum hbc manifesto nec > nec < adeoque $=R$ sit), *tangens* in b ipsius bg per y determinata.

II. Demonstrari potest, esse $\frac{dz^2}{dy^2+bh^2}\sim 1$;

Hinc *limes* ipsius $\frac{dz}{dx}$, et inde z integratione (per x expressum) reperitur. Et potest lineae cuiusvis *in concreto datae* aequatio in S inveniri, e. g. ipsius L. Si enim $\widetilde{am}$ axis ipsius L sit; tum quaevis $\widetilde{cb}$ ex $\widetilde{am}$ secat L (cum (per §.19.) quaevis recta ex a praeter $\widetilde{am}$ ipsum L secet); est vero (si bn axis sit), $X=1:\sin cbn$ (§. 28.), atque $Y=\cot\frac{1}{2}.$ cbn, (§.29) unde fit $Y=X+\sqrt{(X^2-1)}$, seu $e^{\frac{y}{i}}=e^{\frac{x}{i}}+$

$\sqrt{}(e^{\frac{2x}{i}}-1)$ aequatio quaesita. Erit hinc $\frac{dy}{dx} \frown X(X^2-1)^{\frac{-1}{2}}$; atqui $\frac{bh}{dx} = 1 : \sin cbn = X$; adeoque $\frac{dy}{bh} \frown (X^2-1)^{\frac{-1}{2}}$; $1+\frac{dy^2}{bh^2} \frown X^2(X^2-1)^{-1}$, $\frac{dz^2}{bh^2} \frown X^2(X^2-1)^{-1}$, atque $\frac{dz}{dz} \frown X(X^2-1)^{-1}$, $\frac{dz^2}{bh^2} \frown X^2(X^2-1)^{-1}$, atque $\frac{dz}{dx} \frown X^2(X^2-1)^{\frac{-1}{2}}$; unde per integrationem invenitur $z = i(X^2-1)^{\frac{1}{2}} = i \cot cbn$ (uti §. 30.).

III. Manifesto $\frac{du}{d} \frown \frac{hfcbh}{dx}$, quod (nonnisi ab y dependens) iam primum per y exprimendum est; unde u integrando prodit.

Si (Fig. 12.) $ab=p$, $ac=q$, et $cd=r$, atque $cabdc = s$ sit; poterit (uti in II:) ostendi, esse $\frac{ds}{dg} \frown r$, quod $= \frac{1}{2}p\,(e^{\frac{q}{i}} + e^{\frac{-q}{i}})$ atque integrando $s = \frac{1}{2}pi\,(e^{\frac{q}{i}} - e^{\frac{-q}{i}})$. Potest hoc absque integratione quoque deduci. Aequatione e. g, circuli (ex §. 31, III), rectae (ex § 31, II), sectionis coni (per praec) expressis; poterunt areae quoque his lineis clausae exprimi.

Palam est, superficiem t ad figuram planam p (in distantia q) ||*lam*, esse ad p in ratione potentiarum 2*darum* linearum homologarum, sive uti $\frac{1}{4}(e^{\frac{q}{i}} + e^{\frac{-q}{i}})$: 1. Porro computum soliditatis pari modo tractatum. facile patet duas integrationes requirere, (cum et differentiale ipsum hic nonnisi per integrationem determinetur); et ante o.

mnia solidum a p et t ac complexu omnium rectarum ad p ⊥rium fines ipsorum p, t connectentium, clausum quaerendum esse. Reperitur solidum istud (tam per integrationem quam sine ea) $= \frac{1}{2}$ $pi \left[e^{\frac{2q}{i}} - e^{\frac{-2q}{i}}\right] + \frac{1}{2}pq$. Superficies quoque corporum in S determinari possunt, nec non *curvaturae*, *evolutae*, *evolventesque* linearum qualiumvis &c. Quod curvaturam attinet; ea in S aut ipsius L est, aut per radium circuli, aut *distantiam* curvae ad rectam ||lae ab hac recta, determinatur; cum e praecedentibus facile ostendi possit, praeter L, lineas circulares, ac rectae ||las, nullas in plano alias lineas uniformes dari.

IV. Pro circulo est (uti in III.) $\frac{d\bigcirc x}{dx} \frown \bigcirc x$, unde (per §.29.) integrando fit $\odot x = \pi i^2 \left[e^{\frac{x}{i}} - 2 + e^{\frac{-x}{i}}\right]$.

V. Pro area $cabdc = u$ (Fig.9.) (linea Lformi $ab = r$, huic ||la $cd = y$, ac rectis ac, $bd = x$ clausa) est $\frac{du}{dx} \frown y$; atque (§. 24.) $y = re^{\frac{-x}{i}}$; adeoque (integrando) $u = ri\left(1 - e^{\frac{-x}{i}}\right)$. Crescente x in infinitum, fiet in S, $e^{\frac{-x}{i}} \frown o$. adeoque $u \frown ri$. Per *quantitatem* ipsius *mabn*, in posterum limes iste intelligetur. Simili modo invenitur, quod si p sit figura in F: spatium a p et complexu axium e terminis ipsius p ductorum clausum $= \frac{1}{2}$ pi sit.

VI. Si angulus ad centrum segmenti (Fig. 10) sphaerae sit $2u$, peripheria circuli maximi sit p, et arcus fc (∧li u) $= x$; erit $1 : \sin u = p : \bigcirc bc$ (§.25), et hinc $\bigcirc bc = p$ sin u. Interim est $x = \frac{pu}{2\pi}$ ac $dx = \frac{pdu}{2\pi}$. Est porro $\frac{dz}{dx} \frown \bigcirc bc$, et hinc $\frac{dz}{du} \frown \frac{p^2}{2\pi}$ sin u, unde (integrando) $z = \frac{\text{sinv}\, u}{2\pi} p^2$. Cogitetur F in quod p (per meditullium f segmenti transiens) cadit; planis $\widetilde{fem}$, $\widetilde{cem}$ per af, ac ad F $\perp$riter positis, ipsumque in feg, ce secantibus; et considerentur L formis cd (ex c ad feg $\perp$ris) nec non L formis cf; erit $cef = u$ (§.20.), et (§.21.) $\frac{fd}{p} = \frac{\text{sinv}\, u}{2\pi}$, adeoque $z = fd.p$. Ast (§. 21.) $p = \pi.fdg$; itaque $z = \pi.\, fd.fdg$. Est autem (§. 21.) $fd.fdg = fc.fc$; consequ $z = \pi.fc.fc = \bigcirc fc$ in F. Sit iam (Fig.14.) $bj = cj = r$; erit $=$(§. 29.) $2r = i\,(Y - Y^{-1})$, adeoque (§. 21.) $\odot 2r$ (in F) $= \pi i^2\,(Y - Y^{-1})^2$. Est quoque (IV) $\odot 2y = \pi i^2\,(Y^2 - 2 + Y^{-2})$; igitur $\odot 2r$ (in F) $= \odot 2y$, adeoque et $z = \odot 2y$, sive superficies z *segmenti sphaerici aequatur circulo, chorda fc tanquam radio descripto.* Hinc tota sphaerae superficies $= \bigcirc fg = fdg.p = \frac{p^2}{\pi}$, *suntque superficies sphaerarum, uti* 2dae *potentiae peripheriarum earundem maximarum.*

VII. Soliditas sphaerae radii x in S reperitur simili modo $= \frac{1}{2}\pi i^3\,(X^2 - X^{-2}) - 2\pi i^2 x$; superficies per revolutionem lineae cd (Fig.12.) circa ab orta $= \frac{1}{2}\pi i p\,(Q^2 - Q^{-2})$, et corpus per $cabdc$ descriptum $= \frac{1}{2}\pi i^2 p\,(Q^2 + Q^{-2})$. *Quomodo vero omnia a* (IV.) *hucusque tractata, etiam absque integratione perfici possint brevitatis studio supprimitur.*

Demonstrari potest, *omnis expressionis literam i continentis* (adeoque *hypothesi*, quod *detur i*, innixae) *limitem*, *crescente i in infinitum*, *exprimere quantitatem plane pro* Σ (adeoque pro hypothesi *nullius i*), *siquidem non eveniant aequationes identicae.* Cave vero intelligas putari, *systemâ ipsum variari* posse (quod omnino *in se et per se determinatum* est) sed tantum *hypothesin* quod *successive* fieri potest, donec non ad absurdum perducti fuerimus. *Posito* igitur, quod in *tali* expressione litera *i* pro casu, si *S* esset re ipsa, *illam* quantitatem unicam designet, cuius $l = e$ sit; si vero *revera* Σ fuerit, *limes dictus* loco expressionis accipi *cogitetur*: manifesto *omnes* expressiones ex *hypothesi realitatis* ipsius *S* oriundae (hoc sensu) *absolute valent*, etsi *prorsus ignotum sit*, *num* Σ *sit*, *aut non sit.*

Ita e. g. ex expressione in §. 29. obtenta facile (et quidem *tam* differentiationis auxilio, quam *absque* eo) valor notus pro Σ prodit $\bigcirc x = 2\pi x$; ex I. (§. 31.) rite tractato, sequitur $1 : \sin \alpha = c : a$; ex II. vero $\frac{\cos \alpha}{\sin \beta} = 1$, adeoque $\alpha + \beta = R$; aequatio *prima* in III. fit indentica, adeoque *valet* pro Σ, quamvis nihil in eo *determinet*; ex *secunda* autem fluit $c^2 = a^2 + b^2$. *Aequationes notae fundamentales trigonometriae planae* in Σ. Porro inveniuntur (ex §. 32.) pro Σ area et corpus in IV, utrumque $= pq$; ex IV. $\odot x = \pi x^2$; (ex VII) sphaera radii $x = \frac{4}{3} \pi x^3$ &c. Sunt quoque theoremata ad finem (VI) enuntiata manifesto *inconditionate vera.*

§. 33. Superest adhuc quid theoria ista sibi velit, (in §. 32 promissum) exponere.

I. Num Σ aut *S* aliquod *reipsa* sit, indecisum manet.

II. Omnia ex hypothesi *falsitatis* Ax. XI. deducta (semper *sensu* §. 32. intelligendo) *absolute* valent, adeoque *hoc sensu nulli hypothesi innituntur.* Habetur idcirco *trigonometria plana a priori* in qua *solum* systema *ipsum ignotum* adeoque solummodo *absolutae* magnitudines expresionum incognitae manent, per *unicum* vero casum notum, manifesto totum systema figeretur. Trigonometria sphaerica autem in §. 26. absolute stabilitur (Habeturque Geometria, Geometriae planae in Σ prorsus analoga in F).

III. Si *constaret*, Σ esse, nihil hoc respectu amplius incognitum esset; si vero *constaret non esse* Σ, tunc (§. 31.) (e.g.)e lateribus x, y et $\wedge$lo rectilineo ab iis intercepto, in *concreto datis* manifesto in se et per se impossibile esset $\triangle$lum absolute resolvere (i, e.) a priori determinare $\wedge$los ceteros et *rationem lateris tertii* ad duo data; nisi X, Y determinentur, ad quod *in concreto* habere aliquod a oporteret, cuius A notum esset; atque tum i *unitas naturalis longitudinum* esset, (sicuti e est basis logarithmorum naturalium). Si existentia hujus i constiterit; quomodo ad usum saltem quam exactissime construi possit, ostendetur.

IV. Sensu in I et II exposito patet, omnia in spatio methodo recentiorum Analytica (intra justos fines valde laudanda) absolvi posse.

V. Denique lectoribus benevolis haud ingratum futurum est; pro casu illo quodsi non Σ sed S re ipsa esset, circulo aequale rectilineum construi.

§. 34. (Fig.12.) Ex d ducitur $dm|||an$ modo sequenti. Fiat ex d, $db \perp an$; erigatur ,e puncto quovis aliquo a rectae $\overset{\sim}{ab}$, $ac \perp an$ (in dba), et demittatur $de \perp ac$; erit $\bigcirc ed : \bigcirc ab = 1 : \sin z$ (§. 27) siquidem *fuerit* $dm|||bn$. Est vero $\sin z$ non > 1, adeoque ab non $> de$. Descriptus igitur quadrans radio ipsi de aequali, ex a in bac, gaudebit puncto aliquo b vel o cum $\overset{\sim}{bd}$ communi. Priori in casu manifesto $z = R$: in posteriori vero erit (§.2) $(\bigcirc ao = \bigcirc ed) : \bigcirc ab = 1 : \sin aob$, adeoque $z = aoo$. Si itaque fiat $z = aob$; erit $dm|||bn$.

§. 35. (Fig.18.) Si fuerit S reipsa; ducetur recta ad $\wedge$li acuti crus unum $\perp$ris, quae ad alterum $|||$ sit, hoc modo. Sit $am \perp bc$, et accipiatur $ab = ac$ tam parvum (per §.19.), ut si ducatur $bn ||| am$ (§. 34.), sit $abn >$ $\wedge$lo dato. Ducatur porro $cp ||| am$ (§. 34.), fiantque nbq, pcd utrumque $=$ $\wedge$lo dato; et $\widetilde{bq}$, $\widetilde{cd}$ se mutuo secabunt. Secet enim $\widetilde{bq}$ (quod *per constr.* in nbc cadit) ipsam $\widetilde{cp}$ in e; erit (propter $bn \bumpeq cp$) $ebc < ecb$, adeoque $ec < eb$. Sint $ef = ec$, $efr = ecd$, et $fs ||| ep$; cadet fs in bfr. Nam cum $bn ||| cp$, adeoque $bn ||| ep$, atque $bn ||| fs$ sit; erit (§. 14.) $fbn + bfs < (2R = fbn + bfr)$; itaque $bfs < bfr$. Quamobrem $\widetilde{fr}$ secat $\widetilde{ep}$, adeoque $\widetilde{cd}$ quoque ipsam $\widetilde{eq}$ in pnncto aliquo d. Sit iam $dg = dc$, atque $dgt = dcp = gbn$; erit (cum $cd \bumpeq gd$ sit) $bn \bumpeq gt \bumpeq cp$. Si fuerit lineae L formis ipsius bn, punctum in $\widetilde{bq}$ cadens k (§. 19.), et axis kl; erit $bn \bumpeq kl$, adeoque $bkl = bgt = dcp$; sed etiam $kl \bumpeq cp$: cadit ergo k manifesto in g, estque $gt ||| bn$. Si vero ho ipsum bg $\perp$riter bissecet; erit $ho ||| bn$ constructum.

§.36. (Fig.10.) Si fuerint data recta $\widetilde{cp}$ et planum $\widetilde{mab}$, atque fiat $cb \perp \widetilde{mab}$, bn (in $\widetilde{bcp}$) $\perp bc$, et $cq ||| bn$ (§. 34.); sectio ipsius $\widetilde{cp}$ (si haec in bcq cadat) eum $\widetilde{bn}$ (in $\widetilde{cbn}$), adeoque cum $\widetilde{mab}$ reperitur. Et si fuerint data duo plana $\widetilde{pcq}$, $\widetilde{mab}$, et sit $cb \perp \widetilde{mab}$, $cr \perp \widetilde{pcq}$, atque (in $\widetilde{bcr}$) $bn \perp bc$, $cs \perp cr$; cadent bn in $\widetilde{mab}$, et cs in $\widetilde{pcq}$; et sectione ipsarum $\widetilde{bn}$, $\widetilde{cs}$ (si detur) reperta, erit $\perp$ris in pcq per eandem ad $\widetilde{cs}$ ducta, manifesto sesctio ipsorum $\widetilde{mab}$, $\widetilde{pcq}$.

§. 37. (Fig. 7.) In $\widetilde{am} ||| bn$ reperitur tale a, ut sit

$am \simeq bn$; si (per §. 34.) construatur extra $\widetilde{nbm}$, $gt \,|||\, bn$, et fiant $bg \perp gt$, $gc = gb$, atque $cp \,|||\, gt$; ponaturque $\widetilde{tgd}$ ita, ut efficiat cum $\widetilde{tgb}$ $\wedge$lum illi aequalem, quem $\widetilde{pca}$ cum $\widetilde{pcb}$ facit; atque quaeratur (per §. 36.) sectio $\widetilde{dq}$ ipsorum $\widetilde{tgd}$, $\widetilde{nba}$; fiatque $ba \perp dq$. Erit enimvero ob $\triangle$lorum L lineorum in F ipsius bn exortorum similitudinem (§.21.) manifesto $db = da$, et $am \simeq bn$.

Facile hinc patet (L lineis per *solos terminos* datis) reperiri posse etiam *terminos* proportionis 4tum ac medium, atque omnes constructiones geometricas, quae in Σ in plano fiunt, hoc modo in F *absque* XI. *Axiomate* perfici posse. Ita e. g. $4R$ in quotvis partes aequales geometrice dividi potest, si sectionem istam in Σ perficere licet.

§. 38. (Fig.14.) Si construatur (per §. 37.) e. g. $nbq = \frac{1}{3} R$, et fiat (per §. 35) in S ad $\widetilde{bq}$ $\perp$ris $am \,|||\, bn$, atque determinetur (per §. 37.) $jm \simeq bn$; erit, si $ja = x$ sit, (§. 28.) $X = 1 : \sin \frac{1}{3} R = 2$, atque x *geometrice* constructum. Et potest nbq ita computari, ut ja ab i quovis dato minus discrepet, cum nonnisi $\sin nbq = \frac{1}{e}$ esse debeat.

§. 39. (Fig 19.) Si fuerint (in plano) pq et st, ‖ rectae mn (. 27), et ab, cd sint $\perp$res ad mn aequales; manifesto est $\triangle dec \equiv \triangle bea$, adeoque $\wedge$li (forsan mixtilinei) ecp, eat congruent, atque $ec = ea$. Si porro $cf = ag$, erit $\triangle acf \equiv \triangle cag$, et utrumque *quadrilateri fagc* dimidium est. Si *fagc*, *hagk* duo eiusmodi quadrilatera fuerint ad ag, inter pq et st; aequalitas eorum (uti apud *Euclidem*), nec non $\triangle$lorum agc, agh eidem ag insistentium, verticesque in $\widetilde{pq}$ habentium aequalitas patet. Est porro $acf = cag$, $gcq = cga$, atque $acf + acg + gcq = 2R$

(§. 32.), adeoque etiam $cag+acg+cga=2R$: itaque in quovis eiusmodi △lo *acg* summa 3 ∧lorum $=2R$. Sive in *ag* (quae ∥ *mn*) ceciderit autem *recta ag*, sive non; △lorum *rectilineorum agc*, *agh*, *tam ipsorum*, *quam summarum* ∧lorum *ipsorundem*, *aequalitas* in aperto est.

§. 40. (Fig. 20.) *Aequalia* △la *abc*, *abd* (*abhinc rectilinea*) *uno latere aequali gaudentia*, *summas* ∧lorum *aequales habent.* Nam dividat *mn* bifariam tam *ac* quam *bc*, et sit $\widetilde{pq}$ (per *c*) ∥ $\widetilde{mn}$; cadet *d* in $\widetilde{pq}$. Nam si $\widetilde{bd}$ ipsum $\widetilde{mn}$ in puncto *e*, adeoque (§. 39.) ipsum $\widetilde{pq}$ ad distantiam $ef=eb$ secet; erit $\triangle abc=\triangle abf$, adeoque et $\triangle abd=\triangle abf$, unde *d* in *f* cadit: si vero $\widetilde{bd}$ ipsum $\widetilde{mn}$ non secuerit, sit *c* punctum, ubi ⊥lis rectam *ab* bissecans ipsum $\widetilde{pq}$ secat, atque $gs=ht$ ita, ut $\widetilde{st}$ *productam* $\widetilde{bd}$ in puncto aliquo *h* secet (quod fieri posse modo simili patet, ut §. 4.); sint porro $sl=sa$, *lo* ∥ *st*, atque *o* sectio ipsorum $\widetilde{bk}$ et $\widetilde{lo}$: esset tum $\triangle abl=\triangle abo$ (§.39.), adeoque $\triangle abc>\triangle abd$ (contra hyp).

§. 41. (Fig.21.) *Aequalia* △△*abc*, *def*, *aequalibus* ∧lorum *summis gaudent.* Nam secet *mn* tam *ac*, quam *bc*, ita *pq* tam *df* quam *fe* bifariam, et sit *rs* ∥ *mn*, atque *to* ∥ *pq*; erit ⊥ris *ag* ad *rs* aut= ⊥ri *dh* ad *to*, aut altera e. g. *dh* erit maior: in quovis casu ○*df* e centro *a* cum $\widetilde{gs}$ punctum aliquod *k* commune habet, eritque (§. 39.) $\triangle abk=\triangle abc=\triangle def$. Est vero △*akb* (per §. 40.) △lo *dfe*, ac (per §. 39.) △lo *abc aequiangulum.* Sunt igitur etiam △△*abc*, *def* aequiangula.

In *S converti* quoque theorema potest. Sint enim △△*abc*, *def* reciproce aequiangula, atque $\triangle bal=\triangle def$; erit (per praec.) alterum alteri, adeoque etiam △*abc* △lo *abl* aequiangulum, et hinc manifeste $bcl+blc+cbl=2R$. Atqui (ex §. 31.) cuiusvis

$\triangle$li $\wedge$lorum summa in S, est $< 2R$: cadit igitur t in c.

§ 42. (Fig.22.) Si ſuerit *complementum* summae $\wedge$lorum $\triangle$li abc ad $2R$, u, $\triangle$li def vero v; est $\triangle abc$ $\triangle def = u : v$. Nam si quodvis $\triangle$lorum acg, gch, hcb, dfk, kfe sit $= p$, atque $\triangle abc = mp$, $\triangle def = np$; sitque s summa $\wedge$lorum cuiusvis $\triangle$li quod $= p$ est; erit manifesto $2R - u = ms - (m-1)\, 2R = 2R - m\,(2R - s)$, et $u = m\,(2R - s)$, et pariter $v = n\,(2R - s)$. Est igitur $\triangle abc : \triangle def = m : n = u : v$. Ad casum incommensurabilitatis $\triangle$lorum abc, def quoque extendi facile patet.

Eodem modo demonstratur $\triangle$la in superficie sphaerica esse uti *excessus* summarum $\wedge$lorum eorundem supra $2R$. Si 2 $\wedge$li $\triangle$li sphaerici recti fuerint, tertius z erit excessus dictus; est autem $\triangle$ istud (peripheria maxima p dicta) manifesto $= \frac{z}{2\pi} \cdot \frac{p^2}{2\pi}$ (§. 32. VI.); consequ quodvis $\triangle$, cuius $\wedge$lorum excessus $= e$, est $= \frac{zp^2}{4\pi^2}$

§. 43. (Fig.15.) Jam *area* $\triangle$li rectilinei in S per summam $\wedge$lorum exprimetur. Si ab crescat in infinitum: erit (§. 42) $\triangle abc : (R - u - v)$ constans. Est vero $\triangle abc \frown bacn$ (§. 32. V.), et $R - u - v \frown z$ (§. 1.); adeoque $bacn : z = \triangle abc : (R - u - v) = bac'n' : z'$. Est porro manifesto $bdcn : bd'c'n' = r : r' = \text{tang}\, z : \text{tang}\, z$ (§. 30.). Pro $y' \frown o$ autem est $\frac{bd'c'n'}{bac'n'} \frown 1$, nec non $\frac{\text{tang}\, z'}{z'} \frown 1$; consequ. $bdcn : bacn = \text{tang}\, z : z$. Erat vero (§. 32) $bdcn = ri = i^2\, \text{tang}\, z$; est igitur $bacn = zi^2$. Quovis $\triangle$lo cuius $\wedge$lorum summae complementum ad $2R$, z est, in posterum breviter $\triangle$ dicto, erit idcirco $\triangle = zi^2$

Facile hinc liquet, quod si (Fig.14.) $or \parallel am$ et $ro \parallel ab$ fuerint; area inter $\widetilde{or}$, $\widetilde{st}$, $\widetilde{bc}$, comprehensa (quae manifesto limes absolutus est areae triangulorum rectilineorum sine fine crescentium, seu ipsius $\triangle$ pro $z \sim 2R$), sit $= \pi i^2 = \odot i$, in F Limite isto per $\square$ denotato, erit porro (Fig. 15) (per § 30) $\pi r^2 = \text{tang}\, z^2 \square = \odot r$ in F (§ 21) $= \odot s$ (per § 32. VI.), si chorda dc, s dicatur. Si jam radio dato s, circuli in plano (sive radio L formi circuli in F) $\llcorner$riter bisecto, construatur (per § 34) $db \parallel\!\!\!\cong cn$; demissa $\llcorner$ri ca ad db, et erecta $\llcorner$ri cm ad ca; habebitur z; unde (per § 37) $\text{tang}\, z^2$, radio L formi ad lubitum pro unitate assumto, *geometrice determinari potest, per duas lineas uniformes ejusdem curvaturae* (quae solis terminis datis, constructis axibus, manifesto tanquam rectae commensurari, atque hoc respectu rectis aequivalentes spectari possunt).

Porro (Fig. 23) construitur quadrilaterum ex gr. regulare $= \square$, ut sequitur. Sit $abc = R$, $bac = \frac{1}{2}R$, $acb = \frac{1}{4}R$, et $bc = x$; poterit X (ex § 31. II) per meras radices quadraticas exprimi, et (per § 37) construi: habitoque X, (per § 38, sive etiam 29 et 35) x ipsum determinari potest. Estque octuplum $\triangle abc$ manifesto $= \square$, atque *per hoc, circulus planus radii s, per figuram rectilineam, et lineas uniformes ejusdem generis* (*rectis, quoad comparationem inter se, aequivalentes*) *geometrice quadratus; circulus F formis vero eodem modo complanatus: habeturque aut Axioma* XI *Euclidis verum, aut quadratura circuli geometrica*; etsi hucusque indecisum manserit, quodnam ex his duobus revera locum habeat. Quoties tang z^2 *vel* numerus integer *vel* fractio rationalis fuerit, cujus (ad simplicissimam formam reductae) denominator *aut* numerus primus formae 2^m+1 (cujus est etiam $2 = 2^0+1$) *aut* productum fuerit e quotcunque primis hujus formae, quorum (ipsum 2, qui solus quotvis vicibus occurrere potest, excipiendo) quivis *semel* ut factor occurrit: per theoriam polyponorum ill. *GAVSS* (praeclarum nostri imo omnis aevi inventum), etiam ipsi $\tan z^2 \,\square = \odot s$ (et nonnisi pro talibus valoribus ipsius z) figuram rectilineam aequalem constituere licet. Nam *divisio* ipsius $\square$ (theoremate § 42 facile ad quaelibet polygoma extenso) manifesto *sectionem* ipsius 2R requirit, quam (ut ostendi potest) unice sub dicta conditione geometrice perficere licet. In omnibus autem talibus casibus praecedentia facile ad scopum perducent. Et potest quaevis figura rectilinea in polygonum regulare n laterum geometrice converti, siquidem n sub formam *GAVSSianam* cadat.

Superesset denique, (ut res omni numero absolvatur), impossibilitatem, (absque suppositione aliqua) decidendi, num Σ, aut aliquod (et quodnam) S sit, demonstrare: quod tamen occasioni magis idoneae reservatur.

ERRATA.

§. 1. *l.* 6. pro ex $\widetilde{am}$ primum exit, lege, primo non secat $\widetilde{am}$.

§. 4. linea 2 pro $\widetilde{ab}$ lege $\widetilde{ab}$; *l.* 3. lege (per §. 1.), ultima *l.* lege *nam*;

Pag. 4. pro 6 lege § 6: *l.* ult. pro $\widetilde{ba}$ lege $\widetilde{bd}$. Pro bissecare, lege ubique bisecare.

Pag. 5. *l.* 5. a calce, lege $af < ac$; penultima et ult. *l.* lege $\widetilde{ap}$ et $\widetilde{bf}$.

§. 7. Casu 3tio *praemisso* duo priores, adinstar casus 2di §. 10. brevius ac elegantius simul absolvi possunt.

§. 10. a calce, *l.* 4. lege *tgbn*.

§. 11. *l.* 7. et in calce, lege $\widetilde{am}$;

Pag. 9. *l.* 2, pro prortione, lege, extremitate portionis.

§. 17. Demonstrationem ad S restringere haud necesse est; quum facile ita proponatur, ut absolute (pro S et Σ) valeat.

§. 19. penultima *l.* et ult. pro *c* lege *q*.

§. 20. *l.* 2 post 19 claudatur, linea penult. lege, L lineus.

§. 21. *l.* 1. deleatur comma post: in; et *l.* penult. lege $\frac{1}{2} \bigcirc 1$.

§. 22. post Fig. 9. claudatur.

§. 23. *l.* 4. lege, $ab = n.cd$.

§. 24. *l.* 1. lege $Y = X^{\frac{z}{x}}$

Pag. 11. in calce lege $\odot cd$, *l.* penult. lege $\bigcirc ed$.

Pag. 13. *l.* 7. et 8 lege $\frac{\sin u'}{\sin v'} . \bigcirc y'$

Pag. 14. *l.* 4. lege a, c, α; linea 7 lege , pro α.

III. *l.* 3. lege $\frac{\gamma\gamma'}{\alpha\alpha'}$, linea penult. post $e^{\frac{-b}{i}}$ claudatur ; §. 32. deleatur.

Pag. 15. ante §. 32. *l.* penult. duae priores quantitates parenthesibus inclusae quadrari debent, et primus terminus 3tiae exponentem positivum habere. *l.* ult. lege α, β, c

§. 32. I. *l.* 3. a calce, pro hba, lege hbg.

Pag. 16. *l.* 3. lege $\frac{dy}{bh}$, linea 4 lege, atque $\frac{dz}{bh}$; linea 5 lege $X(X^2-1)^{\frac{-1}{2}}$, et dele quod inter duo commata est. III. *l.* 1. lege $\frac{du}{dx}$; *l.* 5. lege $\frac{ds}{dg}$, *l.* 4. a calce, quantitas inclusa quadretur.

Pa. 17. VI. *l.* 1. post segmenti, insere, z:

Pa. 18. linea 12. pro (§ 29). lege (§ 30) ; linea 6. ante VII. dele $z = \odot 2y$, sive ; et VII linea 5, lege $\frac{1}{4}\pi i^2 p(Q - Q^{-1})^2$.

Pag. 19. *l.* 10. lege $= e$; linea 16. lege §. 30 ; linea 13. a calce, lege , III. et in calce, pro ipsum lege, verum.

Pag. 20. *l.* 15. lege *beri* pro *bere* : linea 3. a calce , lege (§. 25.) ; linea 2. lege $= aob$;

Pag. 21. *l.* 2. post *unum*, dele punctum ; et *l.* 3 a. calce pro sesctio, lege sectio.

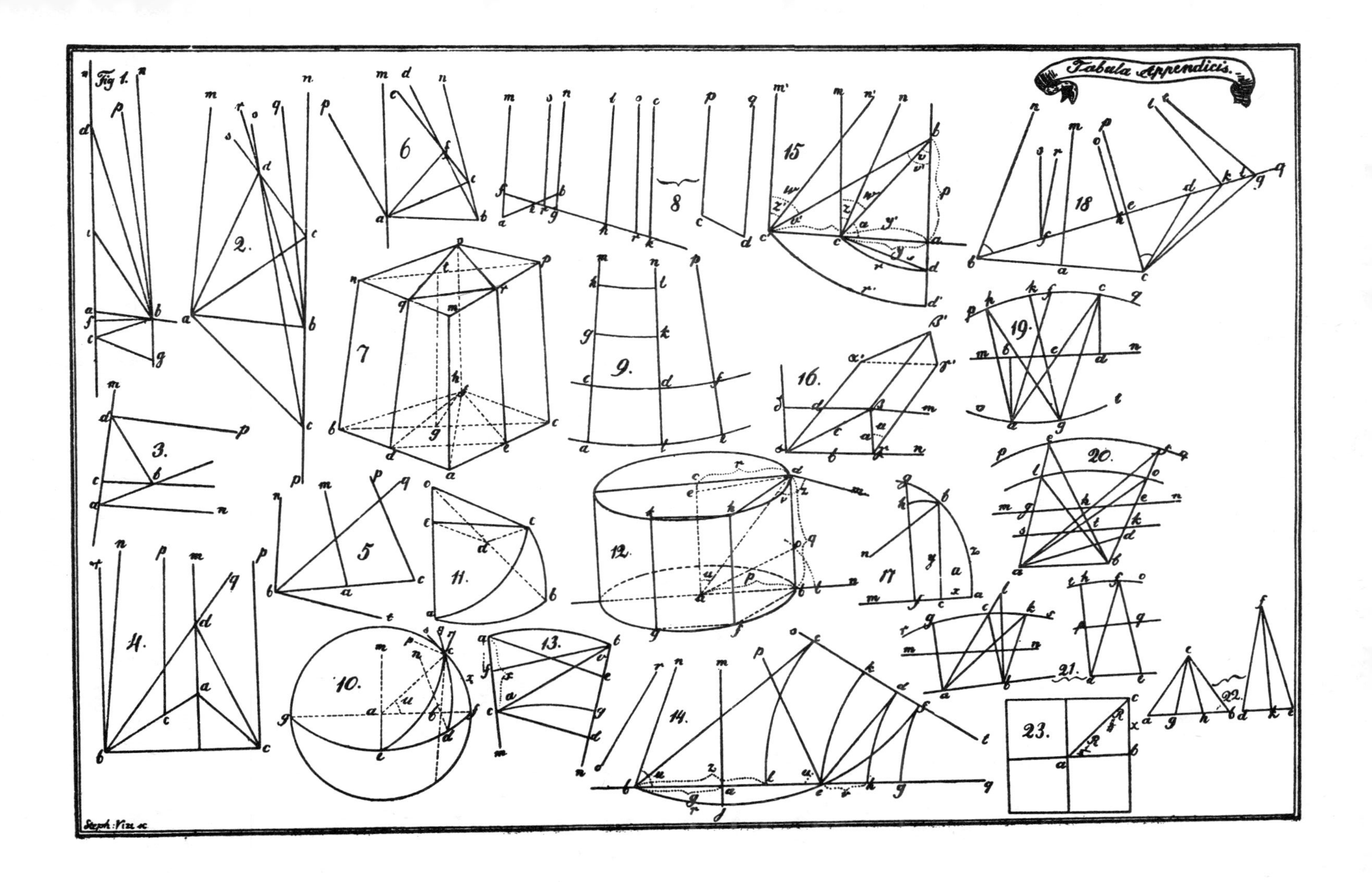
Tabula Appendicis.
Steph: Vizi sc

APPENDIX

THE ABSOLUTELY TRUE SCIENCE OF SPACE

EXPOUNDED INDEPENDENTLY OF THE CORRECTNESS OR FALSENESS (A PRIORI UNDECIDABLE FOR EVER) OF EUCLID'S AXIOM XI: FOR THE CASE OF FALSENESS, WITH A GEOMETRIC QUADRATURE OF THE CIRCLE

BY

JÁNOS BOLYAI

CAPTAIN-ENGINEER OF THE IMPERIAL AND ROYAL AUSTRIAN ARMY

EXPLANATION OF SIGNS

Les us denote by

AB — the collection of all points on the straight line passing through the points A, B;

$\overrightarrow{AB}$ — the half-line (ray) from A through B;

ABC — the collection of all points in the plane containing the non-collinear points A, B, C;

$|AB|C$ — the half-plane having boundary AB and containing C;

$\sphericalangle ABC$ — the smaller of the angular domains bounded by the arms $\overrightarrow{BA}$, $\overrightarrow{BC}$;

$ABCD$ — (if D is an interior point of $\sphericalangle ABC$ and A, B, CD do not intersect each other) the plane domain bounded by $\overrightarrow{BA}$, BC, $\overrightarrow{CD}$ and contained in $\sphericalangle ABC$;

(AB, CD) — (if AB, CD are coplanar and do not intersect) the plane domain bounded by AB and CD;

R — the right angle;

$\equiv$ — congruence*;

$AB \simeq CD$ — the relation $\sphericalangle CAB = \sphericalangle ACD$;

$x \to a$ — that x tends to the limit a;

$\circ r$ — the circumference of the circle of radius r;

$\odot r$ — the area of the circle of radius r.

* There being no risk of confusion, we denote geometric congruence by the symbol the outstanding mathematician GAUSS has used for number-theoretic congruence.

CHAPTER I

PARALLELISM

§1

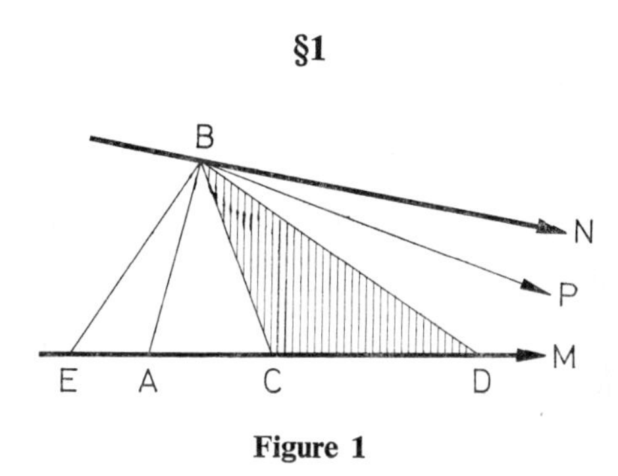

Figure 1

If $\overrightarrow{BN}$ does not, while any other $\overrightarrow{BP}$ in the angular domain ABN does intersect $\overrightarrow{AM}$, we write

$$\overrightarrow{BN} \parallel \overrightarrow{AM}.$$

Obviously, each point B outside the line AM is the origin of one and only one $\overrightarrow{BN}$ with this property. Moreover,

$$\sphericalangle BAM + \sphericalangle ABN \leqq 2R.$$

For, if BC rotates about B until

$$\sphericalangle BAM + \sphericalangle ABC = 2R,$$

there will be a first position where $\overrightarrow{BC}$ does not intersect $\overrightarrow{AM}$, and in this position we have $\overrightarrow{BN} \parallel \overrightarrow{AM}$.

It is also clear that $\overrightarrow{BN} \parallel \overrightarrow{EM}$ for each point E on the line AM (assuming in all such cases that M has been chosen so as to satisfy $\overline{AM} > \overline{AE}$).

If C on AM goes to infinity, and always $\overline{CD} = \overline{BC}$, then

$$\sphericalangle CDB = \sphericalangle CBD < \sphericalangle NBC.$$

But

$$\sphericalangle NBC \to 0.$$

Therefore

$$\sphericalangle ADB \to 0.$$

§2

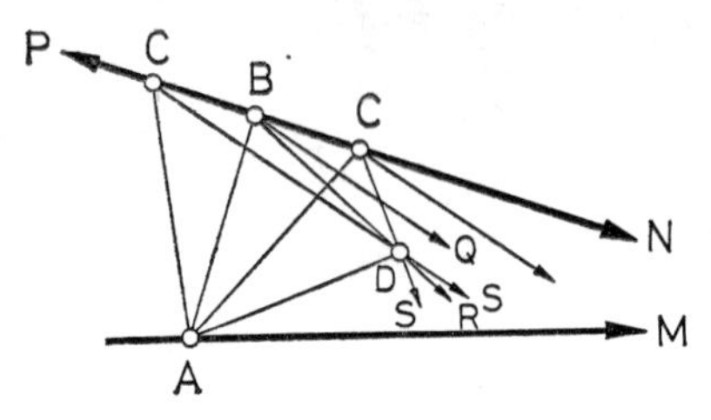

Figure 2

If $\overrightarrow{BN} \| \overrightarrow{AM}$, then also $\overrightarrow{CN} \| \overrightarrow{AM}$.

For let D be an interior point of the plane domain $MACN$. If C is on $\overrightarrow{BN}$, then in view of $\overrightarrow{BN} \| \overrightarrow{AM}$ the half-line $\overrightarrow{BD}$ intersects $\overrightarrow{AM}$ and, therefore, also $\overrightarrow{CD}$ intersects $\overrightarrow{AM}$. If, on the other hand, C is on $\overrightarrow{BP}$, let $\overrightarrow{BQ} \| \overrightarrow{CD}$. By **§1** the half-line $\overrightarrow{BQ}$ lies in the interior of $\sphericalangle ABN$ and intersects $\overrightarrow{AM}$. So $\overrightarrow{CD}$ intersects $\overrightarrow{AM}$.

Thus in both cases any $\overrightarrow{CD}$ in the interior of $\sphericalangle ACN$ does, while $\overrightarrow{CN}$ itself does not intersect $\overrightarrow{AM}$. Consequently, $\overrightarrow{CN} \| \overrightarrow{AM}$.

§3

If $\overrightarrow{BR} \| \overrightarrow{AM}$ and $\overrightarrow{CS} \| \overrightarrow{AM}$, but C is not on the straight line BR, then $\overrightarrow{BR}$ and $\overrightarrow{CS}$ do not intersect each other.

For, assuming that D is a common point of $\overrightarrow{BR}$ and $\overrightarrow{CS}$, **§2** yields $\overrightarrow{DR} \| \overrightarrow{AM}$ and $\overrightarrow{DS} \| \overrightarrow{AM}$. Thus by **§1** the half-lines $\overrightarrow{DR}$ and $\overrightarrow{DS}$ coincide, so that C would be a point of the line BR contrary to the hypothesis.

§4

If $\sphericalangle MAN > \sphericalangle MAB$, then to any point B of $\overrightarrow{AB}$ there can be found a point C on $\overrightarrow{AM}$ such that

$$\sphericalangle BCM = \sphericalangle NAM.$$

For **§1** assures the existence of D such that

$$\sphericalangle BDM > \sphericalangle NAM$$

and, consequently, of P such that

$$\sphericalangle MDP = \sphericalangle MAN;$$

in this case B lies in the plane domain $NADP$. If we shift NAM along $\overrightarrow{AM}$ until $\overrightarrow{AN}$

takes the position $\overrightarrow{DP}$, then $\overrightarrow{AN}$ will at some time pass through B. Therefore necessarily

$$\sphericalangle BCM = \sphericalangle NAM$$

for some point C.

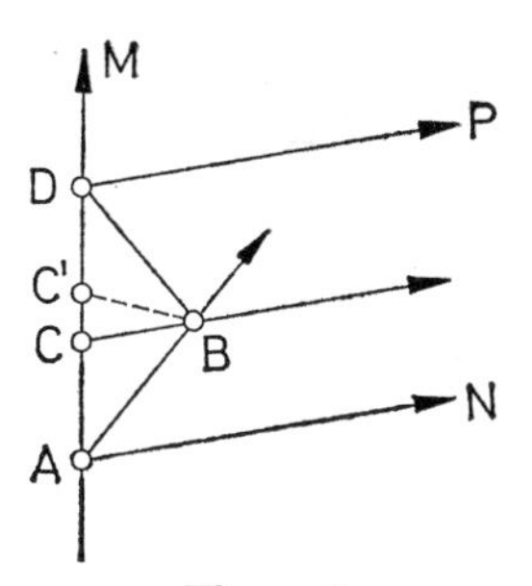

Figure 3

§5

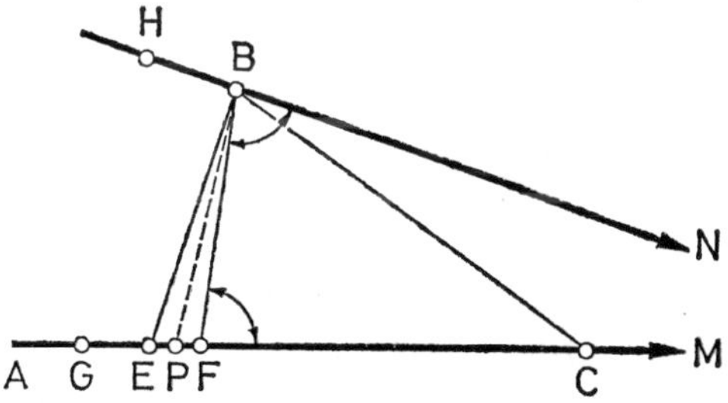

Figure 4

If $\overrightarrow{BN} \parallel \overrightarrow{AM}$, *then there is a point* F *on the line* AM *such that*

$$FM \simeq BN.$$

Really, by **§1** there exists $\sphericalangle BCM$ which is greater than $\sphericalangle CBN$. Further, if $CE = CB$ then

$$EC \simeq BC.$$

Thus

$$\sphericalangle BEM < \sphericalangle EBN.$$

Let the point P run through EC and let $u = \sphericalangle BPM$, $v = \sphericalangle PBN$. It is clear that u is initially smaller but finally greater than the corresponding v. However, u increases from $\sphericalangle BEM$ to $\sphericalangle BCM$ in a *continuous* manner; in fact, by **§4** there is no angle greater than $\sphericalangle BEM$ and smaller than $\sphericalangle BCM$ to which u would not become equal at some time. Similarly, v continuously decreases from $\sphericalangle EBN$ to $\sphericalangle CBN$. So there is a point F on EC having the property

$$\sphericalangle BFM = \sphericalangle FBN.$$

§6

If $\overrightarrow{BN} \| \overrightarrow{AM}$, *whereas G and H are arbitrary points on the lines AM and BN, respectively, then* $\overrightarrow{HN} \| \overrightarrow{GM}$ *and* $\overrightarrow{GM} \| \overrightarrow{HN}$.

For **§1** yields $\overrightarrow{BN} \| \overrightarrow{GM}$ and by **§2** this implies $\overrightarrow{HN} \| \overrightarrow{GM}$.

Further, i. we choose F according to **§5** so that $FM \simeq BN$ then $MFBN = NBFM$. But $\overrightarrow{BN} \| \overrightarrow{FM}$; therefore $\overrightarrow{FM} \| \overrightarrow{BN}$ and by the preceding paragraph $\overrightarrow{GM} \| \overrightarrow{HN}$.

§7

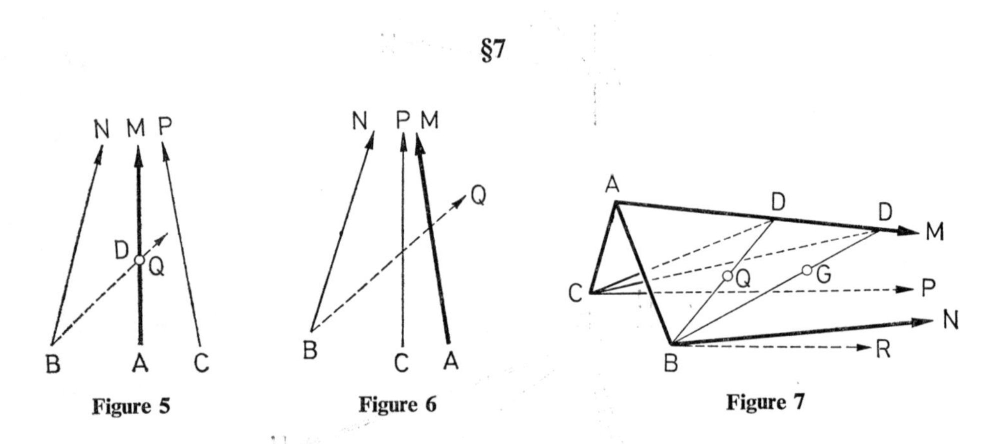

Figure 5 **Figure 6** **Figure 7**

If $\overrightarrow{BN} \| \overrightarrow{AM}$ *as well as* $\overrightarrow{CP} \| \overrightarrow{AM}$ *and C is not on the line BN, then* $\overrightarrow{BN} \| \overrightarrow{CP}$.

In fact, by **§3** the half-lines $\overrightarrow{BN}$ and $\overrightarrow{CP}$ do not intersect each other. $\overrightarrow{AM}$, $\overrightarrow{BN}$ and $\overrightarrow{CP}$ may be either coplanar or not. In the first case $\overrightarrow{AM}$ may be either in the strip (BN, CP) or outside it.

If $\overrightarrow{AM}$, $\overrightarrow{BN}$, $\overrightarrow{CP}$ are coplanar and AM lies in the strip $(BN\ CP)$, then any $\overrightarrow{BQ}$ in $\sphericalangle NBC$ intersects the line AM in some points D, since $\overrightarrow{BN} \| \overrightarrow{AM}$. According to **§6**

$$\overrightarrow{DM} \| \overrightarrow{CP},$$

so that $\overrightarrow{DQ}$ intersects $\overrightarrow{CP}$, and

$$\overrightarrow{BN} \| \overrightarrow{CP}.$$

If however, BN and CP are on the same side of AM, then one of them, say CP, lies between the other two, BN and AM. Therefore any $\overrightarrow{BQ}$ in $\sphericalangle NBA$ intersects $\overrightarrow{AM}$ and, consequently, it intersects $\overrightarrow{CP}$. Thus

$$\overrightarrow{BN} \| \overrightarrow{CP}.$$

If the planes MAB, MAC form an angle, then $\sphericalangle CBN$ has no common point with $\sphericalangle ABN$ outside $\overrightarrow{BN}$, and $\overrightarrow{AM}$ in $\sphericalangle ABN$ has no common point with $\overrightarrow{BN}$. Hence $\sphericalangle NBC$ has no common point with $\overrightarrow{AM}$. However, any half-plane $|BC|Q$ that contains $\overrightarrow{BQ}$, the latter lying in $\sphericalangle NBA$, intersects $\overrightarrow{AM}$. For $\overrightarrow{BN} \| \overrightarrow{AM}$ implies that $\overrightarrow{BQ}$ intersects $\overrightarrow{AM}$ in a point D. Let us turn the half-plane $|BC|D$ about BC until it leaves $\overrightarrow{AM}$ for the first time. Then $|BC|D$ must coincide with $|BC|N$ and, for the same reason, with $|BC|P$. Therefore $\overrightarrow{BN}$ belongs to the half-plane $|BC|P$. Further, if $\overrightarrow{BR} \| \overrightarrow{CP}$ then using the relation $\overrightarrow{AM} \| \overrightarrow{CP}$ and following the line of argument given above it can be seen that $\overrightarrow{BR}$ belongs to the half-plane $|BA|M$. As $\overrightarrow{BR} \| \overrightarrow{CP}$, the half-line $\overrightarrow{BR}$ must belong also to $|BC|P$. Consequently, $\overrightarrow{BR}$ is the intersection of $\sphericalangle MAB$ and $\sphericalangle PCB$. This intersection is therefore identical with $\overrightarrow{BN}$. Hence

$$\overrightarrow{BN} \| \overrightarrow{CP}.$$

Thus if $\overrightarrow{CP} \| \overrightarrow{AM}$ and the point B is outside the plane CAM, then the intersection of $\sphericalangle BAM$ and $\sphericalangle BCP$, that is the half-line $\overrightarrow{BN}$, is parallel to each of $\overrightarrow{AM}$ and $\overrightarrow{CP}$*.

§8

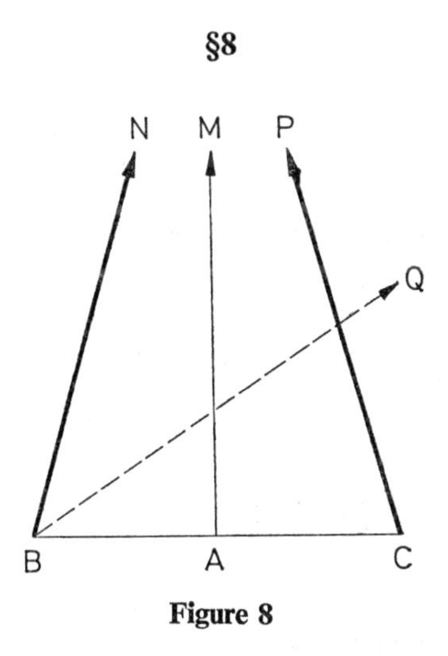

Figure 8

If $\overrightarrow{BN}\|$ and $\simeq \overrightarrow{CP}$, or $\overrightarrow{BN} \| \simeq \overrightarrow{CP}$ for short, and $\overrightarrow{AM}$ in the domain NBCP perpendicularly bisects the distance $\overline{BC}$, then $\overrightarrow{BN} \| \overrightarrow{AM}$.

For if $\overrightarrow{BN}$ intersects $\overrightarrow{AM}$ then, in view of $MABN = MACP$, the half-line $\overrightarrow{CP}$ intersects $\overrightarrow{AM}$ at the same point. The latter would be a common point of $\overrightarrow{BN}$ and $\overrightarrow{CP}$ although $\overrightarrow{BN} \| \overrightarrow{CP}$.

Any $\overrightarrow{BQ}$ in the angular domain CBN intersects $\overrightarrow{CP}$. Therefore $\overrightarrow{BQ}$ intersects $\overrightarrow{AM}$ as well. Consequently,

$$\overrightarrow{BN} \| \overrightarrow{AM}.$$

* In the Errata of the original edition of the *Appendix,* BOLYAI makes the following remark: "If we start from the third case, then the first two can be settled in a shorter and, at the same time, more elegant way on the analogy of the second case occurring in **§10.**"

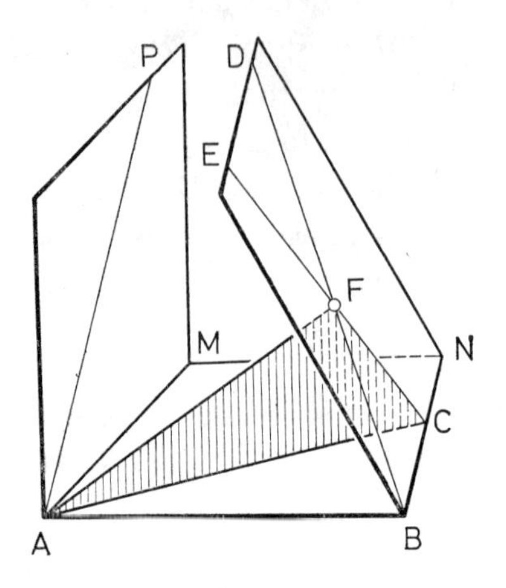

Figure 9

If $\overrightarrow{BN} \| \overrightarrow{AM}$ *and* $MAP \perp MAB$, *while the dihedral angle formed by* $|NB|D$ *and* $|NB|A$ *on that side of the plane* $MABN$ *which contains* $|MA|P$ *is smaller than* R, *then the half-planes* $|MA|P$ *and* $|NB|D$ *intersect each other.*

For let

$$BA \perp AM, \quad AC \perp BN$$

(whether B and C coincide or not) and

$$CE \perp NB$$

in $|NB|D$. Then the hypothesis implies that

$$\sphericalangle ACE < R,$$

and the segment $\overline{AF}$, perpendicular to CE, lies in $\sphericalangle ACE$.

Each of the half-planes $|AB|F$ and $|AM|P$ contains the point A; let their line of intersection be AP. Then

$$\sphericalangle BAP = \sphericalangle BAM = R,$$

since the planes BAM and MAP are perpendicular.

Finally, if $|AB|F$ is turned about AB into $|AB|M$, then $\overrightarrow{AP}$ turns into $\overrightarrow{AM}$. As

$$AC \perp BN \quad \text{and} \quad \overline{AF} < \overline{AC},$$

it is clear that, after the rotation, $\overline{AF}$ will not reach $\overrightarrow{BN}$, and therefore $\overline{BF}$ will fall in the interior of $\sphericalangle ABN$. In this position, $\overrightarrow{BF}$ intersects $\overrightarrow{AP}$ since $\overrightarrow{BN} \| \overrightarrow{AM}$. Thus $\overrightarrow{BF}$ intersects $\overrightarrow{AP}$ in the original position too, and the point of intersection is a common point of $|MA|P$ and $|NB|D$. Consequently, the half-planes $|MA|P$ and $|NB|D$ intersect each other.

Hence it easily follows that *the half-planes $|MA|P$ and $|NB|D$ intersect whenever the sum of the angles they form with the plane domain $MABN$ is $<2R$.*

§10

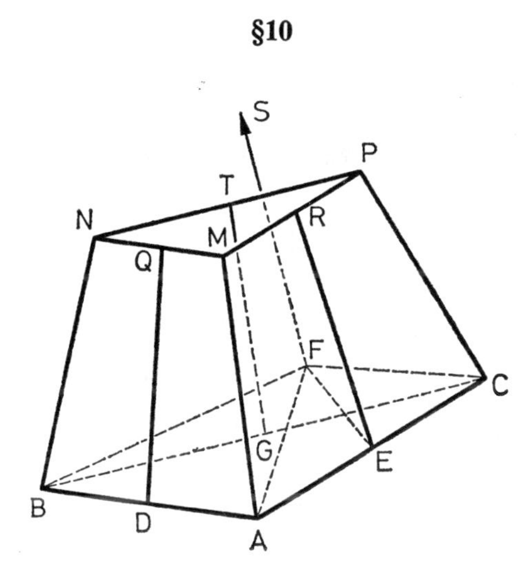

Figure 10

If $\overrightarrow{BN}$ as well as $\overrightarrow{CP}$ are $\| \simeq \overrightarrow{AM}$, then $\overrightarrow{BN} \| \simeq \overrightarrow{CP}$.

In fact, the planes MAB and MAC either form a dihedral angle or coincide.

The first case. Let the plane QDF perpendicularly bisect $\overline{AB}$. Then $DQ \perp AB$, so that $\overrightarrow{DQ} \| \overrightarrow{AM}$ by §8. Similarly, if ERS perpendicularly bisects $\overline{AC}$, then $\overrightarrow{ER} \| \overrightarrow{AM}$. Therefore $\overrightarrow{DQ} \| \overrightarrow{ER}$ by §7. Hence, in view of §9, it follows easily that the planes QDF and ERS intersect each other, and by §7 their line of intersection FS satisfies

$$\overrightarrow{FS} \| \overrightarrow{DQ}.$$

As

$$\overrightarrow{BN} \| \overrightarrow{DQ},$$

also

$$\overrightarrow{FS} \| \overrightarrow{BN}.$$

Further, for any point F on the line FS we have

$$\overline{FB} = \overline{FA} = \overline{FC};$$

thus FS lies in the plane TGF which perpendicularly bisects $\overline{BC}$. But §7 and $\overrightarrow{FS} \| \overrightarrow{BN}$ imply that

$$\overrightarrow{GT} \| \overrightarrow{BN}.$$

6

A similar argument yields

$$\overrightarrow{GT} \parallel \overrightarrow{CP}.$$

Moreover, $\overrightarrow{GT}$ perpendicularly bisects $\overline{BC}$. So by §1

$$TGBN \equiv TGCP$$

and, consequently,

$$\overrightarrow{BN} \parallel \simeq \overrightarrow{CP}.$$

The second case. If $\overrightarrow{BN}$, $\overrightarrow{AM}$ and $\overrightarrow{CP}$ are in one plane, let the half-line $\overrightarrow{FS}$ lying outside this plane satisfy

$$\overrightarrow{FS} \parallel \simeq \overrightarrow{AM}.$$

According to the first case,

$$\overrightarrow{FS} \parallel \simeq \overrightarrow{BN} \quad \text{and} \quad \overrightarrow{FS} \parallel \simeq \overrightarrow{CP}.$$

Therefore

$$\overrightarrow{BN} \parallel \simeq \overrightarrow{CP}.$$

CHAPTER II

THE PARACYCLE AND THE PARASPHERE

§11

Denote by **F** the collection which consists of the point A and all points B such that $\overrightarrow{BN} \parallel \overrightarrow{AM}$ implies $BN \simeq AM$. The intersection of **F** with any plane containing the line AM will be denoted by **L**.

On every line parallel to $\overrightarrow{AM}$ there is one and only one point of **F**. Clearly, $\overrightarrow{AM}$ divides **L** into two congruent parts. We say, $\overrightarrow{AM}$ is an axis of **L**. It is also obvious that in any plane which contains $\overrightarrow{AM}$ there is one and only one **L** having $\overrightarrow{AM}$ for axis. In the plane considered, this **L** will be said to correspond to $\overrightarrow{AM}$. Clearly, if **L** rotates about $\overrightarrow{AM}$, it describes an **F**. We say that $\overrightarrow{AM}$ is an axis of the **F** so obtained and that, in turn, this **F** corresponds to $\overrightarrow{AM}$.

§12

If B is a point of **L**, *where* **L** *corresponds to* $\overrightarrow{AM}$, *and*

$$\overrightarrow{BN} \parallel \simeq \overrightarrow{AM}$$

in accordance with **§11**, *then* **L** *corresponding to* $\overrightarrow{AM}$ *coincides with* **L** *corresponding to* $\overrightarrow{BN}$.

For better distinction, **L** corresponding to $\overrightarrow{BN}$ will be denoted by **l**. Let C be a point of **l** and, in accordance with **§11**,

$$\overrightarrow{CP} \parallel \simeq \overrightarrow{BN}.$$

As $\overrightarrow{BN} \parallel \simeq \overrightarrow{AM}$, **§10** yields

$$\overrightarrow{CP} \parallel \simeq \overrightarrow{AM},$$

so that C lies also on **L**. If, on the other hand, C is a point of **L** and $\overrightarrow{CP} \parallel \simeq \overrightarrow{AM}$, then by **§10**

$$\overrightarrow{CP} \parallel \simeq \overrightarrow{BN}$$

6*

and therefore by **§11** C lies also on **l**. Consequently, **L** and **l** are identical and $\overrightarrow{BN}$ is an axis also for **L**. Thus, all axes of **L** are related by $\parallel \simeq$.

The same property of **F** can be proved in a similar way.

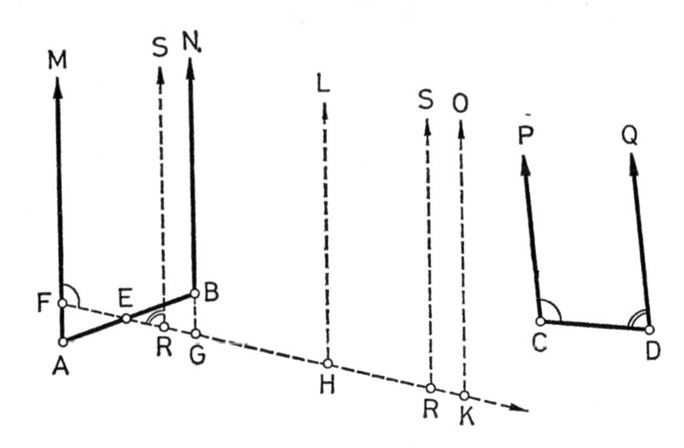

Figure 11

§13

If $\overrightarrow{BN} \parallel \overrightarrow{AM}$, $\overrightarrow{CP} \parallel \overrightarrow{DQ}$ *and* $\sphericalangle BAM + \sphericalangle ABN = 2R$, *then* $\sphericalangle DCP + \sphericalangle CDQ = 2R$.

For let $\overline{EA} = \overline{EB}$ and $\sphericalangle EFM = \sphericalangle DCP$, which can be achieved by **§4**. As

$$\sphericalangle BAM + \sphericalangle ABN = 2R =$$
$$= \sphericalangle ABN + \sphericalangle ABG,$$

we have

$$\sphericalangle EBG = \sphericalangle EAF.$$

Therefore if also $\overline{BG} = \overline{AF}$ then $\Delta EBG \equiv \Delta EAF$ and $\sphericalangle BEG = \sphericalangle AEF$, so that G is on the half-line $\overrightarrow{FE}$. Further

$$\sphericalangle GFM + \sphericalangle FGN = 2R,$$

since $\sphericalangle EGB = \sphericalangle EFA$. Also, in view of §6, $\overrightarrow{GN} \parallel \overrightarrow{FM}$. So if

$$MFRS \equiv PCDQ$$

then by §7 $\overrightarrow{RS} \parallel \overrightarrow{GN}$. Unless $\overline{CD} = FG$, in which case the assertion is obviously true, R lies on either the segment $\overline{FG}$ or its extension.

(I) In the first case, $\sphericalangle FRS$ is not greater than

$$2R - \sphericalangle RFM = \sphericalangle FGN,$$

since $\overrightarrow{RS} \parallel \overrightarrow{FM}$. On the other hand, $\overrightarrow{RS} \parallel \overrightarrow{GN}$ implies that $\sphericalangle FRS$ is not smaller than $\sphericalangle FGN$. Consequently

$$\sphericalangle FRS = \sphericalangle FGN,$$

and

$$\sphericalangle RFM + \sphericalangle FRS = \sphericalangle GFM + \sphericalangle FGN = 2R.$$

Thus

$$\sphericalangle DCP + \sphericalangle CDQ = 2R.$$

(II) If R lies on the extension of $\overline{FG}$, then $\sphericalangle NGR = \sphericalangle MFR$. Let $MFGN \equiv \equiv NGHL \equiv LHKO$ and so on until $\overline{FK} = \overline{FR}$ or $\overline{FK} > \overline{FR}$ for the first time. §7 yields $\overrightarrow{KO} \| \overrightarrow{HL} \| \overrightarrow{FM}$. If K coincides with R, then by §1 $\overrightarrow{KO}$ coincides with $\overrightarrow{RS}$ and therefore

$$\sphericalangle RFM + \sphericalangle FRS = \sphericalangle KFM + \sphericalangle FKO = \sphericalangle KFM + \sphericalangle FGN = 2R.$$

On the other hand, if R lies within $\overline{HK}$, then according to (I) we have

$$\sphericalangle RHL + \sphericalangle HRS = 2R = \sphericalangle RFM + \sphericalangle FRS = \sphericalangle DCP + \sphericalangle CDQ.$$

§14

If $\overrightarrow{BN} \| \overrightarrow{AM}, \overrightarrow{CP} \| \overrightarrow{DQ}$ *and* $\sphericalangle BAM + \sphericalangle ABN < 2R$, *then also* $\sphericalangle DCP + \sphericalangle CDQ < < 2R$.

For, if $\sphericalangle DCP + \sphericalangle CDQ$ were not less than $2R$, then by **§1** it would be equal to $2R$. By **§13** this would imply

$$\sphericalangle BAM + \sphericalangle ABN = 2R$$

contrary to the assumption.

§15

In possession of the results of §§ **13** *and* **14**, *denote by* Σ *the system of geometry based on the hypothesis that* EUCLID's *Axiom XI is true, and denote by* **S** *the system based on the opposite hypothesis.*

All theorems we state without expressly specifying the system Σ *or* **S** *in which the theorem is valid are meant to be absolute, that is, valid independently of whether* Σ *or* **S** *is true in reality.*

§16

If $\overrightarrow{AM}$ *is an axis for some* **L** *then, in system* Σ, *this* **L** *is the straight line perpendicular to* $\overrightarrow{AM}$.

For, if $\overrightarrow{BN}$ is an axis of **L** starting from the point B of **L**, then in Σ

$$\sphericalangle BAM + \sphericalangle ABN = 2\sphericalangle BAM = 2R,$$

so that

$$\sphericalangle BAM = R.$$

If C is any point of the line AB and if $\overrightarrow{CP} \parallel \overrightarrow{AM}$, then by **§13**

$$CP \simeq AM;$$

thus, according to **§11**, C is a point of **L**.

In system **S**, *however, no three points* A, B, C *of* **L** *or* **F** *are on a straight line.*

Really, one of the axes $\overrightarrow{AM}$, $\overrightarrow{BN}$, $\overrightarrow{CP}$, say $\overrightarrow{AM}$ lies between the other two, in which case by **§14** both $\sphericalangle BAM$ and $\sphericalangle CAM$ are smaller than R.

§17

In system **S**, *too,* **L** *is a line and* **F** *is a surface.*

For, in view of **§11**, any plane which is perpendicular to the axis $\overrightarrow{AM}$ and passes through some point of **F** intersects **F** in a circle whose plane, according to **§14**, is not perpendicular to any other axis $\overrightarrow{BN}$.

If we let **F** rotate about the axis $\overrightarrow{BN}$, then by **§12** all points of **F** remain in **F** and the intersection of **F** with any plane not perpendicular to $\overrightarrow{BN}$ describes a surface. By **§12**, however, taking two points A, B of **F** we can place **F** congruently on itself so that A falls on B. Thus **F** is a uniform surface.

Hence by **§§11—12** it follows that **L** is a uniform curve.*

§18

In system **S**, *any plane that passes through the point* A *of* **F** *and stands obliquely to the axis* $\overrightarrow{AM}$ *intersects* **F** *in a circle.*

For let A, B, C be three points of the intersection, and let $\overrightarrow{BN}$ and $\overrightarrow{CP}$ be axes. The planes $AMBN$, $AMCP$ form an angle; otherwise the plane determined by A, B, C would contain, according to **§16**, the axis $\overrightarrow{AM}$, which contradicts the assumption. Therefore by **§10**, the planes perpendicularly bisecting the distances $\overline{AB}$ and $\overline{AC}$ intersect each other in an axis $\overrightarrow{FS}$ of **F**. Thus

$$\overline{FB} = \overline{FA} = \overline{FC}.$$

Further let

$$AH \perp FS.$$

* In the Errata of the original edition, BOLYAI adds: "The proof need not be restricted to system **S**. In fact, it can be easily formulated so as to be valid absolutely (in both systems **S** and Σ)."

Let the plane FAH rotate about FS. Then A describes the circle of radius $\overline{HA}$ passing through B and C, and this circle is in both **F** and the plane ABC. Moreover, by §**16**, **F** and ABC have no common points besides those of the circle $\circ \overline{HA}$.

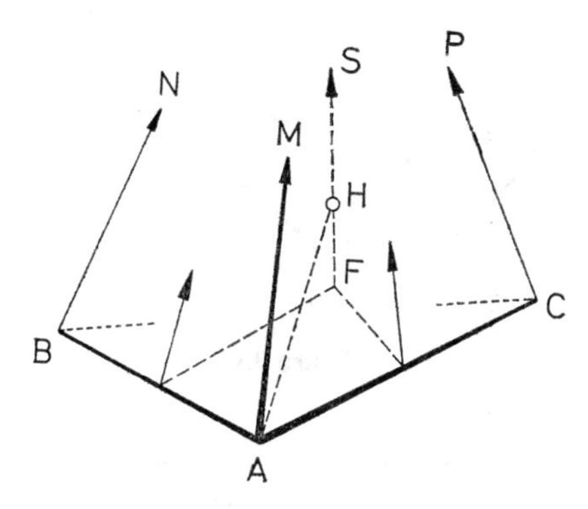

Figure 12

It is also clear that if the arc $\overset{\frown}{FA}$ of an **L** rotates as a radius about the point F within **F** then its endpoint describes the circle $\circ \overline{HA}$.

§19

In system **S**, *a line BT which is perpendicular to the axis $\overrightarrow{BN}$ of some* **L** *and lies in the plane of* **L** *is a tangent to* **L**.

For, by §**14**, B is the only point of $\overrightarrow{BT}$ that lies in **L**. If, however, $\overrightarrow{BQ}$ lies in $\sphericalangle TNB$, then the plane containing BQ and perpendicular to the plane TBN intersects, by §**18**, the **F** corresponding to the axis $\overrightarrow{BN}$ in a circle whose centre lies manifestly on $\overrightarrow{BQ}$. If $\overline{BQ}$ is a diameter of this circle, then $\overrightarrow{BQ}$ obviously intersects the **L** corresponding to the axis $\overrightarrow{BN}$ in the point Q.

§20

According to §§ **11** *and* **18**, *any two points of* **F** *determine an* **L**-*line of* **F**. *As*, however, by §§**16** and **19** **L** is perpendicular to each of its axes, *the angle between any two* **L**-*lines in* **F** *is equal to the dihedral angle between the planes that contain the arms and are perpendicular to* **F**.

§21

Two **L**-*lines $\overrightarrow{AP}$, $\overrightarrow{BQ}$ that lie in the same* **F** *and whose interior angles with a third* **L**-*line AB have a sum less than $2R$ intersect each other*. (In the surface **F**, AP denotes the **L**-line through the points A and P, while $\overrightarrow{AP}$ denotes that half of it which starts from A and contains P.)

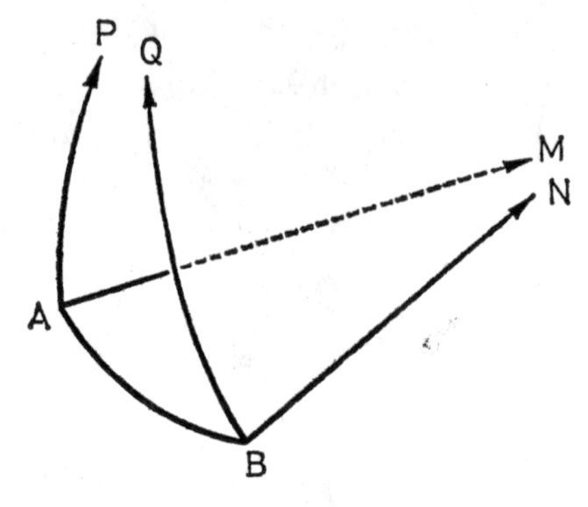

Figure 13

For, if $\overrightarrow{AM}$ and $\overrightarrow{BN}$ are axes of **F**, then $|AM|P$ and $|BN|Q$ intersect each other by §9, and **F** intersects their line of intersection by §§ **7** and **11.** Consequently, also $\overrightarrow{AP}$ and $\overrightarrow{BQ}$ intersect each other.

Hence it is clear that *in the surface* **F**, *if we let* **L**-*lines play the role of straight lines,* Axiom XI as well as the whole geometry and trigonometry of the plane are absolutely valid. Thus, based on the foregoing, trigonometrical functions can be defined just as in system Σ. Also, the circumference of the circle whose radius along an **L**-line of **F** is r equals $2\pi r$ and its area in **F** equals πr^2 (here π denotes half the circumference of the circle of unit radius, that is, the well-known number 3.1415926...).

§22

If AB is an **L**-*line corresponding to the axis* $\overrightarrow{AM}$ *while C lies on* $\overrightarrow{AM}$, *and* $\sphericalangle CAB$ *(whose arms are the axis* $\overrightarrow{AM}$ *and the* **L**-*formed* $\overrightarrow{AB}$*) moves to infinity first along* $\overrightarrow{AB}$ *and afterwards along* $\overrightarrow{BA}$, *then the path CD of C is the* **L**-*line corresponding to the axis* $\overrightarrow{CM}$.

Really, denote by **L′** the **L**-line corresponding to $\overrightarrow{CM}$. Let D be any point of the path CD. Let

$$\overrightarrow{DN} \parallel \overrightarrow{CM},$$

and B a point of **L** on the line DN. Then

$$BN \simeq AM \quad \text{and} \quad \overline{AC} = \overline{BD},$$

whence

$$DN \simeq CM,$$

so that D lies on **L′**. If, however, D lies on **L′**,

$$\overrightarrow{DN} \parallel \overrightarrow{CM}$$

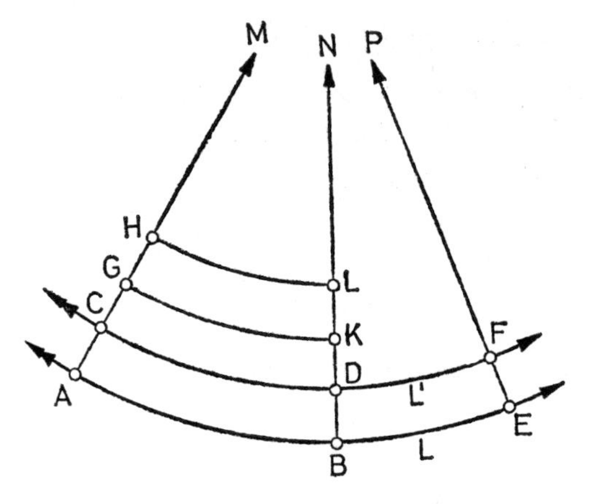

Figure 14

and B is a point of both **L** and DN, then

$$AM \simeq BN \quad \text{and} \quad CM \simeq DN,$$

which obviously imply that

$$\overline{BD} = \overline{AC},$$

D is on the path of C, and **L**′ coincides with CD. For such **L**′, we write

$$\mathbf{L}' \parallel \mathbf{L}.$$

§23

If CDF and ABE are **L**-formed lines such that by **§22**

$$CDF \parallel ABE,$$

further if

$$\widehat{AB} = \widehat{BE},$$

and $\overrightarrow{AM}$, $\overrightarrow{BN}$, $\overrightarrow{EP}$ are axes, then obviously

$$\widehat{CD} = \widehat{DF}.$$

Also, if A, B, E are three arbitrary points of the line AB and

$$\widehat{AB} = n \cdot \widehat{CD},$$

then

$$\widehat{AE} = n \cdot \widehat{CF}$$

and therefore

$$\widehat{AB} : \widehat{CD} = \widehat{AE} : \widehat{CF};$$

the latter relation holds even if $\overset{\frown}{AB}$, $\overset{\frown}{AE}$, $\overset{\frown}{CD}$ are incommensurable. Thus *the ratio* $\overset{\frown}{AB}:\overset{\frown}{CD}$ *is independent of the length of* $\overset{\frown}{AB}$ *and is completely determined by the distance* $\overline{AC}$. *Whenever we denote the length* $\overline{AC}$ *by a small letter* (say x), *the ratio* $\overset{\frown}{AB}:\overset{\frown}{CD}$ *will be denoted by the capital letter of the same name* (X).

§24

For any x and y, with the notation introduced in §23,

$$Y = X^{\frac{y}{x}}.$$

For, either is one of the values x, y a multiple of the other (say, y a multiple of x) or not.

If

$$y = nx,$$

then let

$$x = \overline{AC} = \overline{CG} = \overline{GH}$$

and so on, until

$$\overline{AH} = y.$$

Further let

$$\overset{\frown}{CD} \parallel \overset{\frown}{GK} \parallel \overset{\frown}{HL}.$$

§23 yields

$$X = \overset{\frown}{AB}:\overset{\frown}{CD} = \overset{\frown}{CD}:\overset{\frown}{GK} = \overset{\frown}{GK}:\overset{\frown}{HL}.$$

Hence

$$\frac{\overset{\frown}{AB}}{\overset{\frown}{HL}} = \left(\frac{\overset{\frown}{AB}}{\overset{\frown}{CD}}\right)^n,$$

that is,

$$Y = X^n = X^{\frac{y}{x}}.$$

If x and y are multiples of one and the same z, say

$$x = mz, \quad y = nz,$$

then by the foregoing

$$X = Z^m, \quad Y = Z^n$$

and, consequently,

$$Y = X^{\frac{n}{m}} = X^{\frac{y}{x}}.$$

The result can be easily extended to the case where x and y are incommensurable.

If $q = y - x$, then obviously $Q = Y:X$.

It is also clear that in system Σ for any x we have

$$X = 1.$$

In system **S**, however, we always have

$$X > 1,$$

and for any arcs $\widehat{AB}$, $\widehat{ABE}$ there is a $\widehat{CDF}$ such that

$$\widehat{CDF} \parallel \widehat{ABE} \quad \text{and} \quad \widehat{CDF} = \widehat{AB};$$

hence

$$(AM,\ BN) \equiv (AM, EP),$$

although the latter can be any multiple of the former. However strange this result should be, it obviously still does not prove the absurdity of system **S**.

CHAPTER III

TRIGONOMETRY

§25

In any rectilinear triangle, the circumferences of the circles of radii equal to the sides are to each other as are the sines of the angles opposite to them.

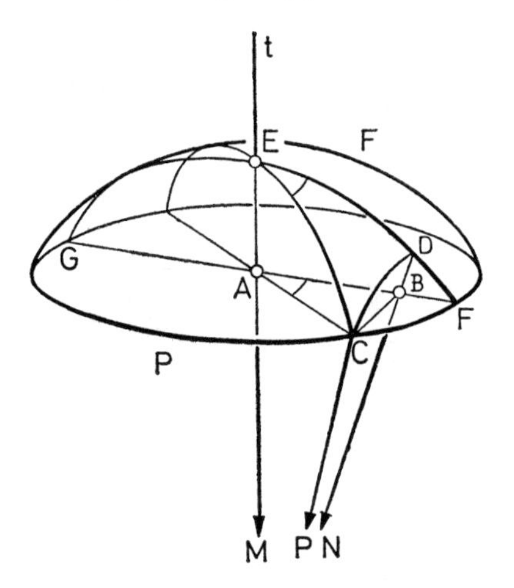

Figure 15

For let

$$\sphericalangle ABC = R, \quad \overrightarrow{AM} \perp BAC$$

and

$$\overrightarrow{BN} \parallel \overrightarrow{AM}, \quad \overrightarrow{CP} \parallel \overrightarrow{AM}.$$

Then the plane CAB is perpendicular to the plane $AMBN$ and, since $CB \perp BA$, also CB is perpendicular to $AMBN$, so that the planes $CPBN$, $AMBN$ are perpendicular to each other. Let the **F** corresponding to the axis $\overrightarrow{CP}$ intersect the lines BN, AM in the points D, E and the strips (CP, BN), (CP, AM), in the **L**-formed arcs $\widehat{CD}$, $\widehat{CE}$, $\widehat{DE}$, (BN, AM) respectively. Then, by **§20**, $\sphericalangle CDE$ will be equal to the dihedral angle between NDC and NDE, that is R, and a similar argument yields

$$\sphericalangle CED = \sphericalangle CAB.$$

According to §**21**, in the triangle *CED* formed of **L**-lines we have*

$$\widehat{EC}:\widehat{DC} = 1:\sin DEC = 1:\sin CAB.$$

Again by §**21**, for the circles drawn on **F**

$$\widehat{EC}:\widehat{DC} = \circ\widehat{EC}:\circ\widehat{DC},$$

and by §**18**

$$\circ\widehat{EC}:\circ\widehat{DC} = \circ\overline{AC}:\circ\overline{BC}.$$

Thus we also have

$$\circ\overline{AC}:\circ\overline{BC} = 1:\sin CAB,$$

from which the validity of the assertion for any triangle follows.

§26

In any spherical triangle, the sines of the sides are to one another as are the sines of the angles opposite to them.

For let

$$\sphericalangle ABC = R,$$

and let the plane *CED* be perpendicular to the radius *OA* of the sphere. Then the plane *CED* is perpendicular to the plane *AOB* and, as also *BOC* is perpendicular to *BOA*,

$$CD \perp OB.$$

However, according to §**25**, in the triangles *CEO* and *CDO* we have

$$\circ\overline{EC}:\circ\overline{OC}:\circ\overline{DC} = \sin\widehat{AC}:1:\sin\widehat{BC} = \sin COE:1:\sin COD,$$

and again by §**25**

$$\circ\overline{EC}:\circ\overline{DC} = \sin CDE:\sin CED.$$

Consequently,

$$\sin\widehat{AC}:\sin\widehat{BC} = \sin CDE:\sin CED.$$

But

$$\sphericalangle CDE = R = \sphericalangle CBA$$

and

$$\sphericalangle CED = \sphericalangle CAB,$$

so that

$$\sin\widehat{AC}:\sin\widehat{BC} = 1:\sin A.$$

Thus the spherical trigonometry which can hence be deduced has obtained a foundation independent of Axiom XI.

* In the Latin original, the obscure insertion "where the radius is always taken to be 1" occurs at this place.

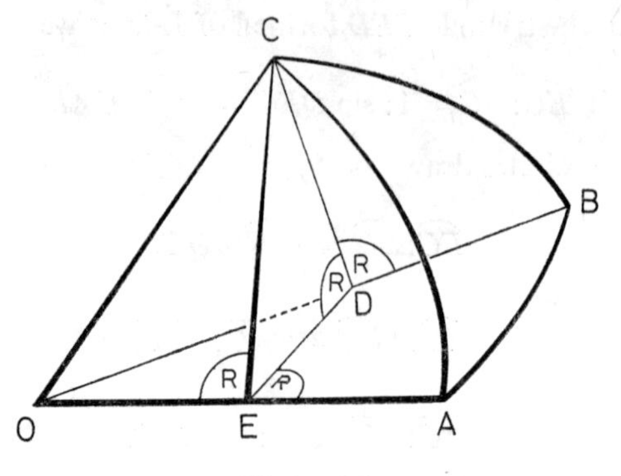

Figure 16

§27

If AC and BD are perpendicular to AB and $\sphericalangle CAB$ *moves along the line AB, then denoting the path of the point C by* $\overset{\frown}{CD}$ *we have*

$$\overset{\frown}{CD} : \overline{AB} = \sin u : \sin v.$$

In fact, let

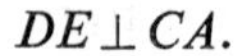

$$DE \perp CA.$$

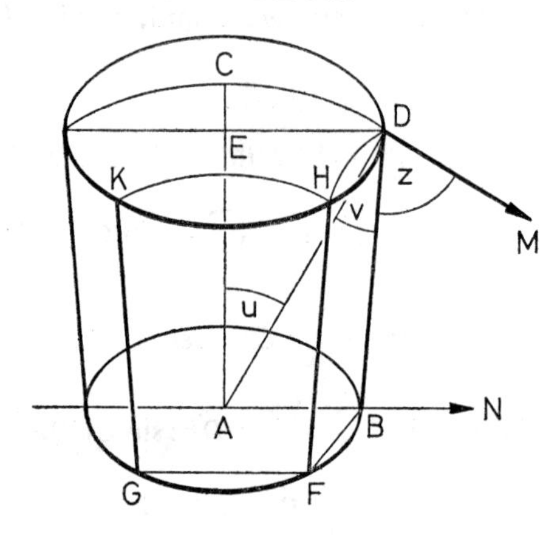

Figure 17

For the triangles ADE, ADB §25 yields

$$\circ\overline{ED} : \circ\overline{AD} : \circ\overline{AB} = \sin u : 1 : \sin v.$$

If the domain $BACD$ rotates about AC, then B and D describe $\circ\overline{AB}$ and $\circ\overline{ED}$, respectively; the surface described by the arc $\overset{\frown}{CD}$ mentioned above will be denoted by $\odot\overset{\frown}{CD}$. Further, let $BFG\ldots$ be a polygon inscribed in $\odot\overline{AB}$. The planes containing the sides BF, FG, ..., respectively, and perpendicular to the plane of $\odot \overline{AB}$ intersect $\odot \widehat{CD}$ in a curvilinear polygon with the same number of sides. Similarly to **§23**, it can be proved that

$$\widehat{CD}:\overline{AB} = \widehat{DH}:\overline{BF} = \widehat{HK}:\overline{FG} = \ldots,$$

which yields

$$(\widehat{DH}+\widehat{HK}+\ldots):(\overline{BF}+\overline{FG}+\ldots) = \widehat{CD}:\overline{AB}.$$

If each of the sides $\overline{BF}$, $\overline{FG}$, ... tends to the limit 0, then obviously

$$\overline{BF}+\overline{FG}+\ldots \to \circ\,\overline{AB}$$

and

$$\widehat{DH}+\widehat{HK}+\ldots \to \circ\,\overline{ED}.$$

Thus, also,

$$\circ\,\overline{ED}:\circ\,\overline{AB} = \widehat{CD}:\overline{AB}.$$

But, as we have just seen,

$$\circ\,\overline{ED}:\circ\,\overline{AB} = \sin u:\sin v.$$

Consequently,

$$\widehat{CD}:\overline{AB} = \sin u:\sin v.$$

If we remove AC from BD to infinity, then the ratio $\widehat{CD}:\overline{AB}$, that is $\sin u:\sin v$, remains constant. By **§1**, however,

$$u \to R,$$

and if $\overrightarrow{DM} \| \overrightarrow{BN}$ then

$$v \to z.$$

Hence

$$\widehat{CD}:\overline{AB} = 1:\sin z.$$

For the path $\widehat{CD}$ in question we write

$$\widehat{CD} \| \overline{AB}.$$

§28

If $\overrightarrow{BN} \| \simeq \overrightarrow{AM}$ *and* C *lies on* $\overrightarrow{AM}$, *whereas* $\overline{AC}=x$, *then with the notation of* **§23** *we have*

$$X = \sin u:\sin v.$$

For, if CD and AE are perpendicular to BN, and BF is perpendicular to AM, then it can be seen as in **§27** that

$$\circ\,\overline{BF}:\circ\,\overline{CD} = \sin u:\sin v.$$

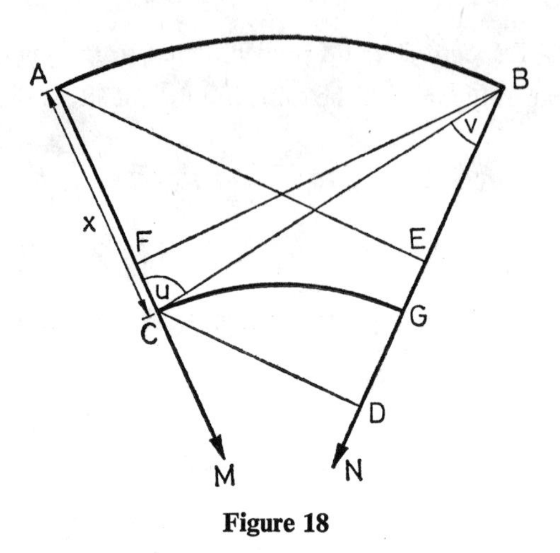

Figure 18

But, obviously, $\overline{BF}=\overline{AE}$. So

$$\circ\overline{EA}:\circ\overline{DC}=\sin u:\sin v.$$

On the other hand, in the **F**-surfaces corresponding to $\overrightarrow{AM}$ and $\overrightarrow{CM}$ (that intersect the strip (AM, BN) in the arcs $\widehat{AB}$ and $\widehat{CG}$) **§21** yields

$$\circ\overline{EA}:\circ\overline{DC}=\widehat{AB}:\widehat{CG}=X.$$

As a result,

$$X=\sin u:\sin v.$$

§29

If $\sphericalangle BAM=R$, $\overline{AB}=y$ *and* $\overrightarrow{BN}\|\overrightarrow{AM}$, *then in system* **S**

$$Y=\operatorname{ctg}\frac{1}{2}u.$$

For, if

$$\overline{AB}=\overline{AC}\quad\text{and}\quad\overrightarrow{CP}\|\overrightarrow{AM}$$

so that

$$BN\simeq CP,$$

further if

$$\sphericalangle PCD=\sphericalangle QCD,$$

then by **§19** there is a $\overrightarrow{DS}$ perpendicular to $\overrightarrow{CD}$ and parallel to $\overrightarrow{CP}$. Hence by **§1**

$$\overrightarrow{DT}\|\overrightarrow{CQ}.$$

Moreover, let BE be perpendicular to $\overrightarrow{DS}$. **§7** yields $\overrightarrow{DS}\|\overrightarrow{BN}$; thus by **§6** $\overrightarrow{BN}\|\overrightarrow{ES}$.

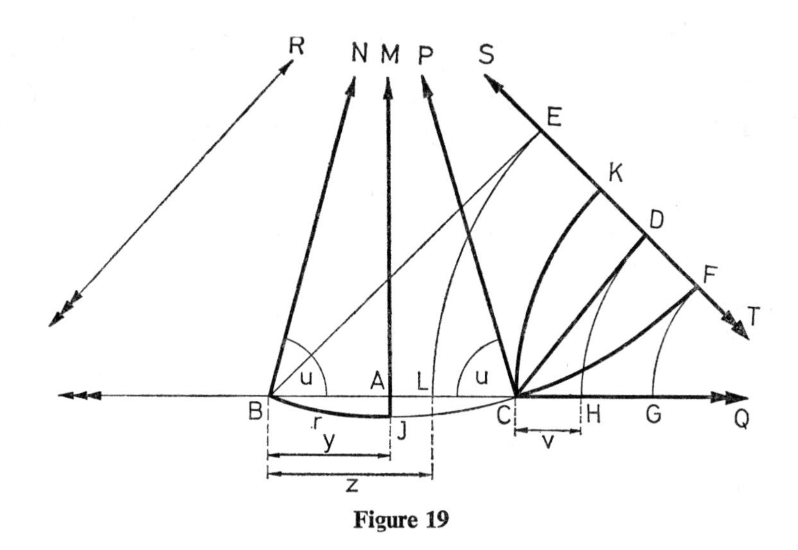

Figure 19

As, however, $\overrightarrow{DT} \| \overrightarrow{CQ}$, we have $\overrightarrow{BQ} \| \overrightarrow{ET}$. Consequently, in view of **§1**,

$$\sphericalangle EBN = \sphericalangle EBQ.$$

Let the arc $\widehat{BCF}$ be contained in the **L**-line corresponding to the axis $\overrightarrow{BN}$, and let $\widehat{FG}$, $\widehat{DH}$, $\widehat{CK}$ and $\widehat{EL}$ be some arcs of the **L**-lines corresponding to the axes $\overrightarrow{FT}$, $\overrightarrow{DT}$, $\overrightarrow{CQ}$ and $\overrightarrow{ET}$, respectively. From **§22** it is clear that

$$\overline{HG} = \overline{DF} = \overline{DK} = \overline{HC}.$$

Hence

$$\overline{CG} = 2 \cdot \overline{CH} = 2v.$$

The relation

$$\overline{BG} = 2 \cdot \overline{BL} = 2z$$

can be verified similarly. But

$$\overline{BC} = \overline{BG} - \overline{CG}.$$

Therefore

$$y = z - v.$$

Consequently, according to **§24**,

$$Y = Z : V.$$

Finally, in view of **§28**,

$$Z = 1 : \sin \frac{1}{2} u$$

and

$$V = 1 : \sin \left(R - \frac{1}{2} u \right).$$

Thus

$$Y = \operatorname{ctg} \frac{1}{2} u.$$

§30

Making use of §25 it is easy to see that, for solving the problem of plane trigonometry in system S, the expression for the circumference of the circle in terms of the radius is needed. This expression, in turn, can be obtained by the rectification of the **L**-line.

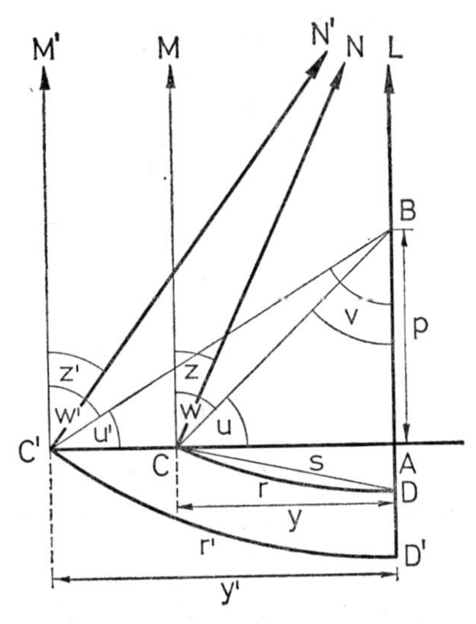

Figure 20

Let $\overrightarrow{AB}$, $\overrightarrow{CM}$, $\overrightarrow{C'M'}$ be perpendicular to $\overrightarrow{AC}$, and let B be any point of $\overrightarrow{AB}$. By §25

$$\sin u : \sin v = \circ p : \circ y$$

and

$$\sin u' : \sin v' = \circ p : \circ y'.$$

Consequently,

$$\frac{\sin u}{\sin v} \cdot \circ y = \frac{\sin u'}{\sin v'} \cdot \circ y'.$$

But according to §27

$$\sin v : \sin v' = \cos u : \cos u',$$

so that

$$\frac{\sin u}{\cos u} \cdot \circ y = \frac{\sin u'}{\cos u'} \cdot \circ y',$$

or

$$\circ y : \circ y' = \operatorname{tg} u' : \operatorname{tg} u = \operatorname{tg} w : \operatorname{tg} w'.$$

Moreover, let

$$\overrightarrow{CN} \parallel \overrightarrow{AB}, \quad \overrightarrow{C'N'} \parallel \overrightarrow{AB},$$

and let $\widehat{CD}$, $\widehat{C'D'}$ be **L**-lines perpendicularly intersecting the straight line AB. By §21 we also have

$$\circ y : \circ y' = r : r';$$

thus

$$r:r' = \text{tg } w:\text{tg } w'.$$

Now let the distance p measured from A increase indefinitely. Then

$$w \to z \quad \text{and} \quad w' \to z'.$$

Hence also

$$r:r' = \text{tg } z:\text{tg } z'.$$

The constant $r:\text{tg } z$ which is independent of r will be denoted by k. If $y \to 0$, then

$$\frac{r}{y} = \frac{k \cdot \text{tg } z}{y} \to 1$$

and therefore

$$\frac{y}{\text{tg } z} \to k.$$

§29 yields

$$\text{tg } z = \frac{1}{2}(Y - Y^{-1}).$$

Thus

$$\frac{2y}{Y - Y^{-1}} \to k$$

or, by **§24**,

$$\frac{2yK^{\frac{Y}{K}}}{K^{\frac{2y}{k}} - 1} \to k.$$

But it is well known that, for $y \to 0$ the limit of this expression is

$$\frac{k}{\log K}.$$

Consequently,

$$\frac{k}{\log K} = k$$

and

$$K = e = 2.7182818 \ldots.$$

Here, too, this number seems to have outstanding significance. Thus if, from now onwards, k denotes the distance whose corresponding K is just equal to e, then

$$r = k \cdot \text{tg } z.$$

On the other hand, in **§21** we pointed out that

$$\circ y = 2\pi r.$$

So by **§24**

$$\circ y = 2\pi k \cdot \text{tg } z = \pi k(Y - Y^{-1}) =$$

$$= \pi k\left(e^{\frac{y}{k}} - e^{-\frac{y}{k}}\right) = \frac{\pi y}{\log Y}(Y - Y^{-1}).$$

§31

In system **S**, the knowledge of three equations is sufficient for solving every rectilinear right triangle, and this renders easy to solve any triangle. Namely, we only need to know the equations that specify the relation

(I) between a, c, α,

(II) between a, α, β,

(III) between a, b, c,

where a, b denote the legs, c the hypotenuse, and α, β the angles opposite to the legs. From them, of course, the three remaining equations can be deduced by elimination.

(I) According to §§ **25** and **30**

$$1:\sin\alpha=(C-C^{-1}):(A-A^{-1})=\left(e^{\frac{c}{k}}-e^{-\frac{c}{k}}\right):\left(e^{\frac{a}{k}}-e^{-\frac{a}{k}}\right).$$

This is the equation for a, c, α.

(II) If $\overrightarrow{BM}\|\overrightarrow{CN}$, then §**27** yields

$$\cos\alpha:\sin\beta=1:\sin u,$$

while §**29** yields

$$1:\sin u=\frac{1}{2}(A+A^{-1}).$$

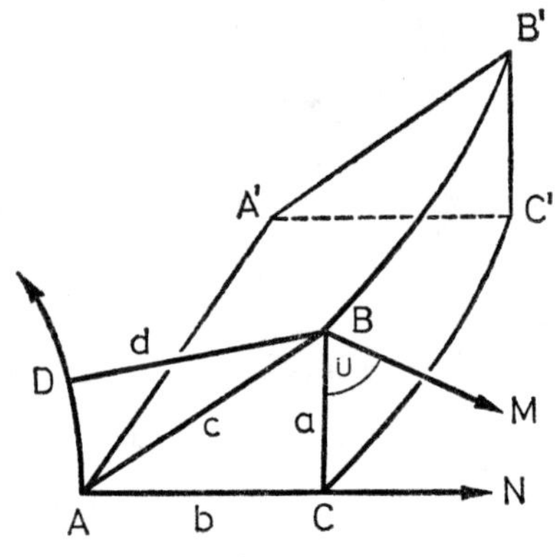

Figure 21

Consequently,

$$\cos\alpha:\sin\beta=\frac{1}{2}(A+A^{-1})=\frac{1}{2}\left(e^{\frac{a}{k}}+e^{-\frac{a}{k}}\right).$$

This is the equation for a, α, β.

(III) If AA' is perpendicular to the plane BAC, if (with the notation introduced in §27) $\widehat{BB'}\|\overline{AA'}$ and $\widehat{CC'}\|\overline{AA'}$, finally if the plane $B'A'C'$ is perpendicular to the line AA', then in the same way as in §**27** it follows that

$$\frac{\widehat{BB'}}{\widehat{CC'}}=\frac{1}{\sin u}=\frac{1}{2}(A+A^{-1}),$$

$$\frac{\widehat{CC'}}{\overline{AA'}}=\frac{1}{2}(B+B^{-1}),$$

and

$$\frac{\widehat{BB'}}{\overline{AA'}} = \frac{1}{2}(C+C^{-1}).$$

Hence

$$\frac{1}{2}(C+C^{-1}) = \frac{1}{2}(A+A^{-1}) \cdot \frac{1}{2}(B+B^{-1})$$

and, consequently,

$$(e^{\frac{c}{k}}+e^{-\frac{c}{k}}) = \frac{1}{2}(e^{\frac{a}{k}}+e^{-\frac{a}{k}})(e^{\frac{b}{k}}+e^{-\frac{b}{k}}).$$

This is the equation for a, b, c.

If

$$\sphericalangle CAD = R$$

and

$$BD \perp AD,$$

then

$$\circ c : \circ a = 1 : \sin\alpha$$

and

$$\circ c : \circ d = 1 : \cos\alpha,$$

where $d = \overline{BD}$. Therefore, denoting $\circ x \cdot \circ x$ by $\circ x^2$, obviously

$$\circ a^2 + \circ d^2 = \circ c^2.$$

But, in view of §27 and (II),

$$\circ d = \circ b \cdot \frac{1}{2}(A+A^{-1}),$$

which yields

$$(e^{\frac{c}{k}}-e^{-\frac{c}{k}})^2 = \frac{1}{4}(e^{\frac{a}{k}}+e^{-\frac{a}{k}})^2(e^{\frac{b}{k}}-e^{-\frac{b}{k}})^2+(e^{\frac{a}{k}}-e^{-\frac{a}{k}})^2.$$

This is another equation for a, b, c; its right-hand side can be easily brought to symmetric form.

Finally, from

$$\frac{\cos\alpha}{\sin\beta} = \frac{1}{2}(A+A^{-1})$$

and

$$\frac{\cos\beta}{\sin\alpha} = \frac{1}{2}(B+B^{-1})$$

by the aid of (III) we obtain

$$\operatorname{ctg}\alpha \cdot \operatorname{ctg}\beta = \frac{1}{2}(e^{\frac{c}{k}}+e^{-\frac{c}{k}}).$$

This is the equation for α, β, c.

CHAPTER IV

APPLICATION OF THE METHODS OF ANALYSIS RELATION BETWEEN GEOMETRY AND REALITY

§32

It remains to indicate the way of solving problems in system S and finally, this being accomplished for some examples most often encountered, to describe frankly what our theory offers.

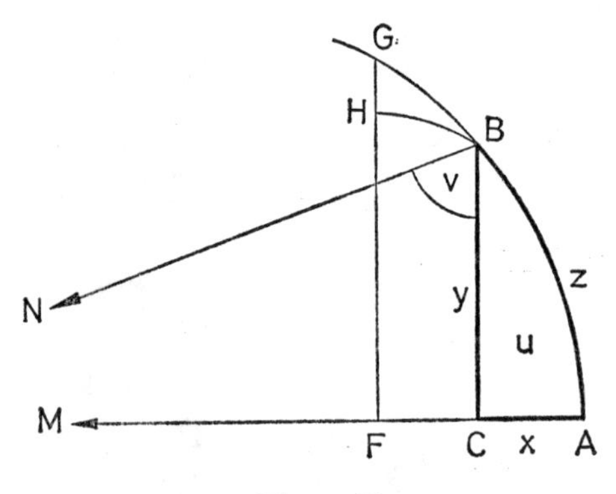

Figure 22

(I) Let $\widehat{AB}$ be a plane curve whose equation in rectangular coordinates reads $y=f(x)$. Denote by dz any increment of z, and by dx, dy, du the increments, corresponding to dz, of x, y and the area u, respectively. Further, with the notation introduced in §27, let $\widehat{BH} \| \overline{CF}$. Making use of §31, express $\widehat{BH}/dx$ as a function of y. Find the limit of dy/dx as dx tends to the limit 0 (which will be tacitly assumed whenever taking limits of this kind). From these, the limit of $dy/\widehat{BH}$ and, consequently, the value tg HBG become known. Since $\sphericalangle HBC$ is neither greater nor smaller than the right angle, it is a right angle. As a result, we obtain the tangent of the curve $\widehat{BG}$ at the point B in terms of y.

(II) It can be proved that

$$\frac{dz^2}{dy^2+\widehat{BH}^2} \to 1.$$

Hence we can calculate the limit of dz/dx and find z as a function of x by integration.

In system **S**, it is possible to deduce the equation of any curve given concretely; for instance, that of the **L**-line.

If $\overrightarrow{AM}$ is an axis of **L**, then any half-line $\overrightarrow{CB}$ starting from the point C of $\overrightarrow{AM}$ intersects **L**, since by **§19** every straight line through A, excepting AM, intersects **L**. If, however, $\overrightarrow{BN}$ is also an axis, then by **§28**

$$X = 1 : \sin CBN$$

and by **§29**

$$Y = \operatorname{ctg} \frac{1}{2} CBN.$$

Hence

$$Y = X + \sqrt{X^2 - 1}$$

or

$$e^{\frac{y}{k}} = e^{\frac{x}{k}} + \sqrt{e^{\frac{2x}{k}} - 1}.$$

This is the equation required. It yields

$$\frac{dy}{dx} \to X(X^2-1)^{-\frac{1}{2}}.$$

But

$$\frac{\widehat{BH}}{dx} \to 1 : \sin CBN = X.$$

Thus

$$\frac{dy}{\widehat{BH}} \to (X^2-1)^{-\frac{1}{2}},$$

$$1 + \left(\frac{dy}{\widehat{BH}}\right)^2 \to X^2(X^2-1)^{-1},$$

$$\left(\frac{dz}{\widehat{BH}}\right)^2 \to X^2(X^2-1)^{-1},$$

$$\frac{dz}{\widehat{BH}} \to X(X^2-1)^{-\frac{1}{2}}$$

and

$$\frac{dz}{dx} \to X^2(X^2-1)^{-\frac{1}{2}}.$$

By integration we obtain

$$z = k(X^2-1)^{\frac{1}{2}} = k \operatorname{ctg} CBN$$

in accordance with **§30**.

(III) For the area of the domain $HFCBH$ we obviously have

$$\frac{du}{dx} \to \frac{HFCBH}{dx}.$$

This depends only on y and must first be expressed in terms of y; then u is obtained by integration.

Denoting (see Fig. 17) $\overline{AB}$ by p, $\overline{AC}$ by q, $\widehat{CD}$ by r and the area of the domain $CABDC$ by s, it can be shown as in (II) that

$$\frac{ds}{dq} \to r = \frac{1}{2} p\left(e^{\frac{q}{k}} + e^{-\frac{q}{k}}\right),$$

and hence by integration

$$s = \frac{1}{2} pk\left(e^{\frac{q}{k}} - e^{-\frac{q}{k}}\right).$$

This can be derived also without integration.

If we deduce the equation of the circle using §31, (III), or that of the straight line using §31, (II), or that of a conic section using the arguments above, then the areas enclosed by these lines can also be calculated.

Clearly, the area of the surface t parallel to the plane figure p at distance* q is to p as are the second powers of the homologous line segments or, more explicitly, as

$$\frac{1}{4}\left(e^{\frac{q}{k}} + e^{-\frac{q}{k}}\right)^2 : 1.$$

It is easy to see that calculating the volume in a similar way requires two integrations (here, in fact, even the differential can only be obtained by integration), and that first of all the volume of the solid enclosed by p, t and all lines perpendicular to the plane of p and connecting the boundaries of p and t must be determined. We find (whether by means of integration or without it) that this volume equals

$$\frac{1}{8} pk\left(e^{\frac{2q}{k}} - e^{-\frac{2q}{k}}\right) + \frac{1}{2} pq.$$

In **S**, also the surface area of a solid can be obtained, as well as the curvature, evolute, evolvent of an arbitrary curve, etc. As to the curvature in system **S**, it may either be that of an **L**-line, or characterised by the radius of a circle or by the distance between a curve which is parallel to a straight line and that line. Really, in view of the foregoing it is easy to show that there are no uniform plane curves other than **L**-lines, circles, and curves parallel to a straight line.

(IV) As in (III), for the area and circumference of the circle we have

$$\frac{d(\odot x)}{dx} \to \circ x.$$

Owing to §30, integration yields

$$\odot x = \pi k^2\left(e^{\frac{x}{k}} - 2 + e^{-\frac{x}{k}}\right).$$

* Thus the surface in question is a region of a distance surface. *(Editor's remark.)*

(V) For the area* $CABDC=u$ enclosed by the L-line $\widehat{AB}=r$, by the L-line $\widehat{CD}=y$ parallel to it, and by the segments $\overline{AC}=\overline{BD}=x$ we have

$$\frac{du}{dx}\to y.$$

From §24

$$y=re^{-\frac{x}{k}},$$

and hence, by integration,

$$u=rk\left(1-e^{-\frac{x}{k}}\right).$$

If $x\to\infty$, then $e^{-\frac{x}{k}}\to 0$, so that in **S**

$$u\to rk.$$

In what follows, the size of the domain $MABN$ will always mean this limit.

It can be established in a similar way that if p is a figure in **F** then the volume enclosed by p and all axes starting from the boundary of p equals $\frac{1}{2}pk$.

(VI)** If the central angle of the spherical cap z is $2u$, the circumference of the great circle is p, and the circular arc corresponding to central angle u is $\widehat{FC}=x$, then by §25

$$1:\sin u=p:\circ\overline{BC}.$$

Thus

$$\circ\overline{BC}=p\sin u.$$

On the other hand,

$$x=\frac{pu}{2\pi}\quad\text{and}\quad dx=\frac{p\,du}{2\pi}.$$

Further

$$\frac{dz}{dx}\to\circ\overline{BC},$$

so that

$$\frac{dz}{du}\to\frac{p^2}{2\pi}\sin u,$$

and by integration

$$z=\frac{1-\cos u}{2\pi}p^2.$$

Consider the **F**-surface that contains the circle p passing through the centre F of the spherical cap. The line*** t intersects **F** perpendicularly at the point E. The planes FEM and CEM containing the radii $\overline{AF}$ and $\overline{AC}$ are perpendicular to the surface **F** and

* See Fig. 14.

** See: Tabula Appendices Fig 10 *(Editor's remark)*.

*** See Fig. 15.

intersect it in the curves FEG and CE. Furthermore, consider the **L**-arc $\overset{\frown}{CD}$ starting from C and perpendicular to the curve FEG, and the **L**-arc $\overset{\frown}{CF}$.

According to **§20** we have

$$\sphericalangle CEF = u$$

and by **§21**

$$\frac{\overset{\frown}{FD}}{p} = \frac{1-\cos u}{2\pi}.$$

Therefore

$$z = \overset{\frown}{FD} \cdot p.$$

But from **§21** it follows that

$$p = \pi \cdot \overset{\frown}{FG}.$$

Thus

$$z = \pi \cdot \overset{\frown}{FD} \cdot \overset{\frown}{FG}.$$

On the other hand, again by **§21**,

$$\overset{\frown}{FD} \cdot \overset{\frown}{FG} = \overset{\frown}{FC}^2.$$

Consequently, for the circular domain of the **F**-surface we have

$$z = \pi \cdot \overset{\frown}{FC}^2 = \odot\, \overset{\frown}{FC}.$$

Now let* $\overset{\frown}{BJ} = \overset{\frown}{CJ} = r$. By **§30**

$$2r = k(Y - Y^{-1}).$$

Consequently, according to **§21**, for the circular domain of **F**

$$\odot\, 2r = \pi k^2 (Y - Y^{-1})^2.$$

On the other hand, (IV) yields

$$\odot\, 2y = \pi k^2 (Y^2 - 2 + Y^{-2}).$$

Thus for the circular domain of **F** we have

$$\odot\, 2r = \odot\, 2y.$$

Therefore *the area of the spherical cap z is equal to the area of the circle described with the chord $\overline{FC}$ as radius.*

Hence the surface area of the whole sphere is

$$\odot\, \overline{FG} = \overset{\frown}{FG} \cdot p = \frac{p^2}{\pi}.$$

Therefore *the surface areas of two spheres are to each other as are the second powers of the circumferences of their great circles.*

* See Fig. 19.

(VII) We find in a similar way that in system S the volume of the sphere of radius x is

$$\frac{1}{2}\pi k^3(X^2 - X^{-2}) - 2\pi k^2 x.$$

The area of the surface obtained by rotation of the arc* $\widehat{CD}$ about $\overline{AB}$ is

$$\frac{1}{2}\pi kp(Q^2 - Q^{-2}),$$

whereas the volume of the solid described by the figure $CABDC$ equals

$$\frac{1}{4}\pi k^2 p(Q - Q^{-1})^2.$$

For the sake of brevity, we shall not explain how everything we have presented beginning from (IV) *can be deduced also without integration.*

It can be proved that *if* k *tends to infinity then the limit of any expression containing* k (hence based on the hypothesis that k exists) *coincides with the value of the same quantity valid in* Σ (which system involves the non-existence of any k) *unless an identity is obtained.* Beware of the impression, however, as if *the system itself could be altered*; it is completely *determined in itself and by itself.* Only the *hypothesis* can be altered, as far as we are not led to a contradiction. Thus, *assuming* that in such expressions k denotes the unique value whose K is equal to e if system S is true in reality, while the expression is thought to be replaced by the limit mentioned above if system Σ is actually valid, then it is clear that all expressions obtained from the hypothesis of the reality of S are, in this sense, absolutely valid, though it remains perfectly unknown whether Σ is, or is not, fulfilled.

Thus, for instance, the expression obtained in **§30** yields (either by or without differentiation) the value

$$\circ x = 2\pi x,$$

well known is system Σ. From §31, (I) in the usual way we deduce

$$1 : \sin \alpha = c : a.$$

Moreover, from (II),

$$\frac{\cos \alpha}{\sin \beta} = 1,$$

that is

$$\alpha + \beta = R.$$

The first equation of (III) becomes an identity, hence is valid in Σ, though does not determine anything in it; the second, however, implies

$$c^2 = a^2 + b^2.$$

These are the well-known fundamental equations of plane trigonometry in system Σ.

* See Fig. 17.

Further, according to **§32**, in system Σ the area as well as the volume appearing in (III) are equal to

$$pq,$$

(IV) yields

$$\odot x = \pi x^2,$$

from (VII) it follows that the volume of the sphere of radius x is

$$\frac{4}{3}\pi x^3,$$

etc. Obviously, the theorems announced at the end of (VI) are also unconditionally true.

§33

It remains to make clear the implications of our theory, as promised in **§32**.

(I) It rests undecided whether system Σ or some system **S** is true in actual fact.

(II) Everything we have deduced from the hypothesis that Axiom XI is false is absolutely valid (always in the sense of **§32**) and therefore, in this sense, does not lean on any assumption. So, there is an a priori plane trigonometry in which only the true system is unknown, hence only the absolute magnitudes of the expressions remain undetermined, but on the basis of a single known case, obviously, the whole system could be fixed. On the other hand, spherical trigonometry was established in **§26** in an absolute way. Further, on the surface **F** we have a geometry completely analogous to the plane geometry of system Σ.

(III) If we knew that Σ is valid, there would remain no open question in this respect. On the other hand, if we knew that Σ is not valid, then (see **§31**) starting e.g. from the sides x, y and the angle between them, each concretely given, it would obviously be impossible in itself and by itself to resolve the triangle in an absolute manner, i.e. a priori determine the remaining angles and the ratio of the third side to both of the given ones; this could only be done by determining X and Y, for which purpose we would need a concrete value of a whose corresponding A is known. In the latter case, k would be a natural unit of length (just as e is the base of natural logarithms). Assuming the existence of k, we shall show how it can be constructed, at least for practical use, as accurately as possible.

(IV) In the sense of (II) and (III), all spatial problems can apparently be settled by a recent method of analysis which deserves high appreciation if applied within proper limits.

(V) Finally, there comes something not at all disagreeable to the gentle reader: assuming that S, and not Σ, is valid in reality, we construct a rectilinear figure of area equal to that of a circle.

CHAPTER V

CONSTRUCTIONS

§34

From the point D, a half-line $\overrightarrow{DM}$ parallel to $\overrightarrow{AN}$ can be drawn in the following way.

Let DB be the perpendicular from D to AN. At any point A of the line AB, draw AE, the perpendicular to AN in the plane ABD. Further, let

$$DE \perp AE.$$

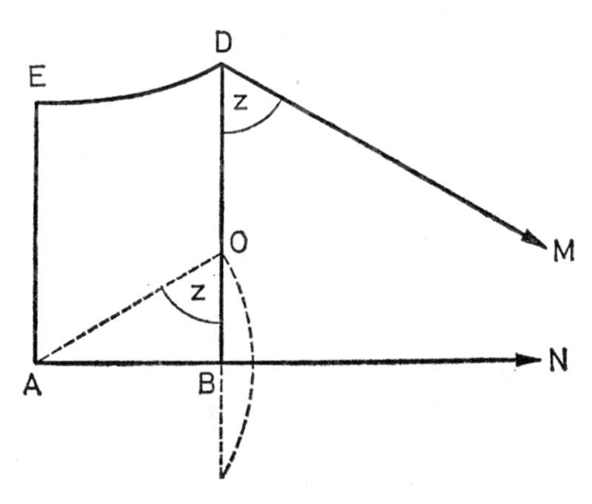

Figure 23

Now, if

$$\overrightarrow{DM} \parallel \overrightarrow{BN},$$

then by **§27**

$$\circ \overline{ED} : \circ \overline{AB} = 1 : \sin z.$$

But $\sin z \leqq 1$, so that $\overline{AB} \leqq \overline{DE}$. Therefore the quadrant of radius $\overline{DE}$ drawn from the centre A in the angular domain BAE has a point B or O in common with the half-line $\overrightarrow{BD}$. In the first case evidently

$$z = R.$$

In the second case, by **§25**,

$$(\circ \overline{AO} = \circ \overline{ED}) : \circ \overline{AB} = 1 : \sin AOB;$$

hence

$$z = \sphericalangle AOB.$$

Thus, if we take the angle z to be equal to $\sphericalangle AOB$ then

$$\overrightarrow{DM} \parallel \overrightarrow{BN}.$$

§35

If **S** is valid in reality, *a straight line that is perpendicular to one arm of an acute angle and parallel to the other can be constructed as follows.*

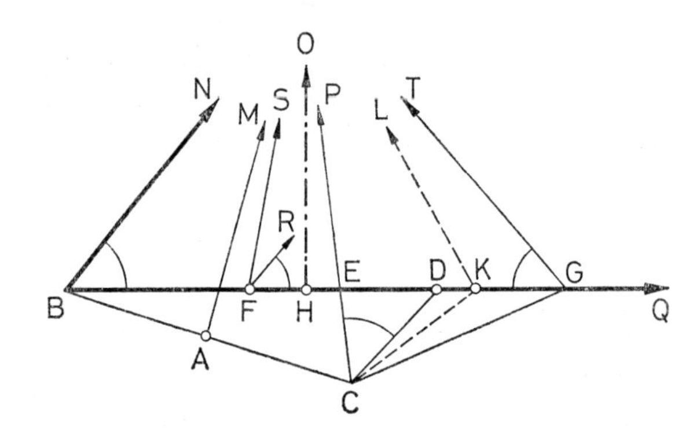

Figure 24

Let $AM \perp BC$. Choose the distance $\overline{AB} = \overline{AC}$ to be so small (see §**19**) that if constructing (by the aid of §**34**) the half-line $\overrightarrow{BN}$ parallel to $\overrightarrow{AM}$, the angle ABN be greater than the angle originally given. Further, making use of §**34** once more, draw the half-line $\overrightarrow{CP}$ that is parallel to $\overrightarrow{AM}$, and let each of $\sphericalangle NBQ$ and $\sphericalangle PCD$ be equal to the given angle. The half-lines $\overrightarrow{BQ}$ and $\overrightarrow{CD}$ intersect each other.

For let $\overrightarrow{BQ}$ (which lies in $\sphericalangle NBC$ by construction) intersect $\overrightarrow{CP}$ at the point E. Since $\overrightarrow{BN} \parallel \simeq \overrightarrow{CP}$, we have

$$\sphericalangle EBC < \sphericalangle ECB$$

and, consequently,

$$\overline{EC} < \overline{EB}.$$

Let

$$\overline{EF} = \overline{EC}, \quad \sphericalangle EFR = \sphericalangle ECD, \quad \text{and} \quad \overrightarrow{FS} \parallel \overrightarrow{EP}.$$

Then $\overrightarrow{FS}$ must fall into $\sphericalangle BFR$. Indeed, as

$$\overrightarrow{BN} \parallel \overrightarrow{CP}$$

so that

$$\overrightarrow{BN} \parallel \overrightarrow{EP} \quad \text{and} \quad \overrightarrow{BN} \parallel \overrightarrow{FS},$$

by **§14** we have

$$\sphericalangle FBN + \sphericalangle BFS < 2R = \sphericalangle FBN + \sphericalangle BFR$$

and, consequently,

$$\sphericalangle BFS < \sphericalangle BFR.$$

Thus $\overrightarrow{FR}$ intersects $\overrightarrow{EP}$, and therefore also $\overrightarrow{CD}$ intersects $\overrightarrow{EQ}$ at some point D.

Now let

$$\overline{DG} = \overline{DC}$$

and

$$\sphericalangle DGT = \sphericalangle DCP = \sphericalangle GBN.$$

Since $CD \simeq GD$, it follows that

$$BN \simeq GT \simeq CP.$$

If K is the point on $\overrightarrow{BQ}$ (see **§19**) and $\overrightarrow{KL}$ is an axis of the **L**-line that corresponds to the axis $\overrightarrow{BN}$, then

$$BN \simeq KL.$$

Hence

$$\sphericalangle BKL = \sphericalangle BGT = \sphericalangle DCP$$

and, on the other hand,

$$KL \simeq CP.$$

So it is obvious that K coincides with G and

$$\overrightarrow{GT} \parallel \overrightarrow{BN}.$$

If, however, $\overrightarrow{HO}$ perpendicularly bisects $\overline{BG}$, then the half-line $\overrightarrow{HO}$ parallel to $\overrightarrow{BN}$ is constructed.

§36

Suppose there is given a half-line $\overrightarrow{CP}$ and a plane MAB. Let CB be perpendicular to MAB, BN be perpendicular to BC in the plane BCP, and $\overrightarrow{CQ}$ be parallel to $\overrightarrow{BN}$ (see **§34**). If $\overrightarrow{CP}$ lies in $\sphericalangle BCQ$ then, in the plane CBN, the point of intersection of $\overrightarrow{CP}$ with $\overrightarrow{BN}$, and hence with MAB, can be found.

Next suppose we are given two planes PCQ and MAB. Let

$$CB \perp MAB, \quad CR \perp PCQ,$$

and in the plane BCR

$$BN \perp BC, \quad CS \perp CR.$$

Then BN and CS lie in MAB and PCQ, respectively. Taking the point of intersection of BN and CS (if it exists), the perpendicular at this point to CS in PCQ will apparently be the intersection of MAB and PCQ.

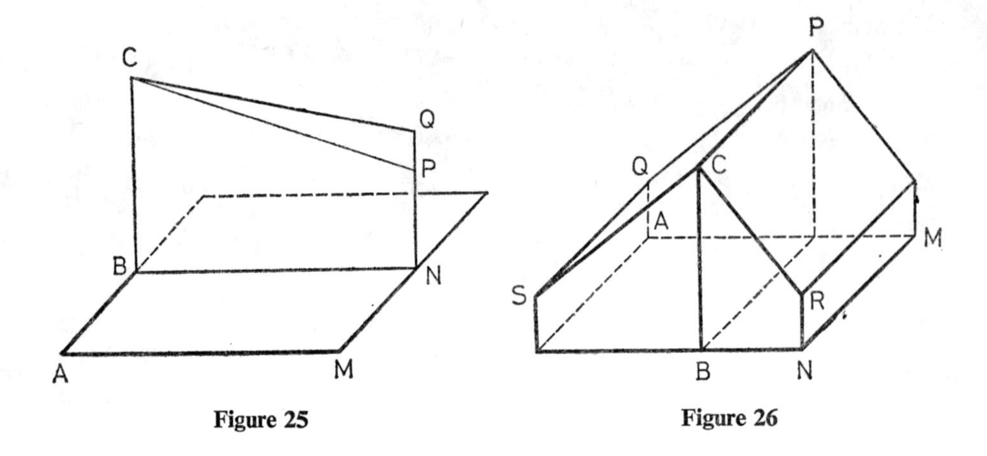

Figure 25

Figure 26

§37

On a half-line $\overrightarrow{AM}$, which is parallel to $\overrightarrow{BN}$, the point A can be chosen so that $AM \simeq BN$.

For let us construct (see §**34**) a half-line $\overrightarrow{GT}$ that is parallel to $\overrightarrow{BN}$ and not in the plane NBM. Let

$$BG \perp GT, \quad \overline{GC} = \overline{GB}, \quad \text{and} \quad \overrightarrow{CP} \parallel \overrightarrow{GT}.$$

Choose the half-plane $|TG|D$ so as to form an angle with $|TG|B$ equal to the angle between $|PC|A$ and $|PC|B$. With the help of §**36**, determine the intersection DQ of the half-planes $|TG|D$ and $|NB|A$. Finally, let

$$BA \perp DQ.$$

Then the triangles formed by the **L**-lines which arise in the **F**-surface corresponding to $\overrightarrow{BN}$ are similar by §**21**. Therefore obviously

$$\overline{DB} = \overline{DA} \quad \text{and} \quad AM \simeq BN.$$

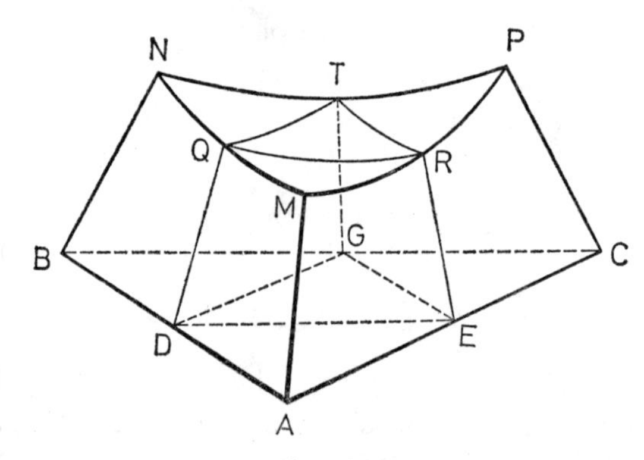

Figure 27

Hence it is easy to see that, *for* **L**-*lines given only by their endpoints, the endpoints of the fourth proportional and the geometric mean can be obtained, and all constructions which are, in system* Σ, *possible in the plane can so, without Axiom XI, be performed in the surface* **F**.

Thus, for instance, $4R$ can be divided into any number of equal parts by geometric construction if that division can be accomplished in system Σ.

§38

Let us construct, say, the angle $\sphericalangle NBQ = \frac{1}{3}R$ with the help of §37. Let $\overrightarrow{AM}$ be perpendicular to $\overrightarrow{BQ}$ and parallel to $\overrightarrow{BN}$ in system **S** (see §35). Determine J by §37 so that $JM \simeq BN$. Then for the distance $\overline{JA} = x$ §28 yields

$$X = 1 : \sin\frac{1}{3}R = 2.$$

Thus x is geometrically constructed.

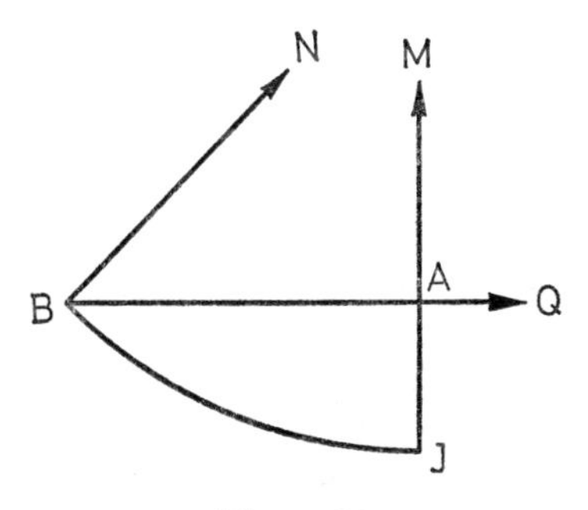

Figure 28

The angle NBQ can be computed so that the difference between $\overline{JA}$ and k be arbitrarily small. Indeed, it is sufficient to assure the validity of the relation

$$\sin NBQ = \frac{1}{e}.$$

§39

If, in the plane, the curves $\widehat{PQ}$ and $\widehat{ST}$ are parallel to the straight line MN (see §27), and the distances $\overline{AB}$ and $\overline{CD}$ are perpendicular to MN and equal to each other, then evidently

$$\triangle DEC \equiv \triangle BEA.$$

8

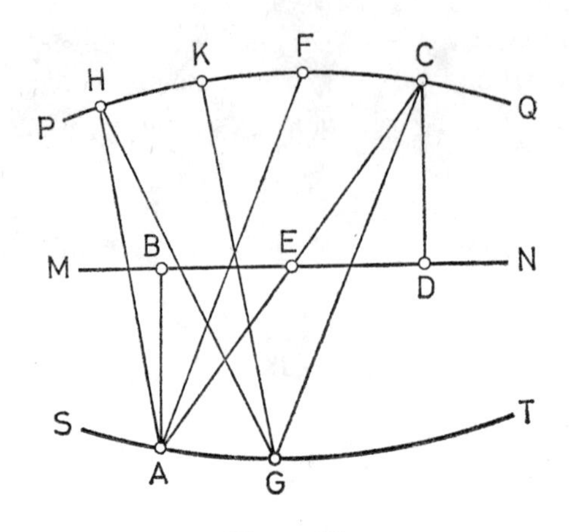

Figure 29

Therefore the angular domains $\sphericalangle ECP$ and $\sphericalangle EAT$ (which may be bounded by lines of mixed kind) are congruent, and

$$\overline{EC} = \overline{EA}.$$

Further, if

$$\widehat{CF} = \widehat{AG}$$

then

$$\triangle ACF \equiv \triangle CAG$$

and each of these triangles is half of the quadrilateral $FAGC$. If $FAGC$ and $HAGK$ are two quadrilaterals of this type based on $\widehat{AG}$ and lying between $\widehat{PQ}$ and $\widehat{ST}$, then (following EUCLID) it can be seen that their areas are equal as are those of the triangles AGC and AGH erected on the same arc $\widehat{AG}$ and having one vertex on $\widehat{PQ}$.

Moreover,

$$\sphericalangle ACF = \sphericalangle CAG, \quad \sphericalangle GCQ = \sphericalangle CGA,$$

and by §32

$$\sphericalangle ACF + \sphericalangle ACG + \sphericalangle GCQ = 2R.$$

Thus

$$\sphericalangle CAG + \sphericalangle ACG + \sphericalangle CGA = 2R.$$

Consequently, in any such triangle ACG the sum of the three angles is $2R$.

Whether the straight line AG is in the curve $\widehat{AG}$ (the latter being parallel to MN) or not, it is now clear that the areas as well as the angle sums of the rectilinear triangles AGC, AGH are equal.

§40

Two (from now on, rectilinear) triangles ABC and ABD of equal area and with one side equal have equal angle sums.

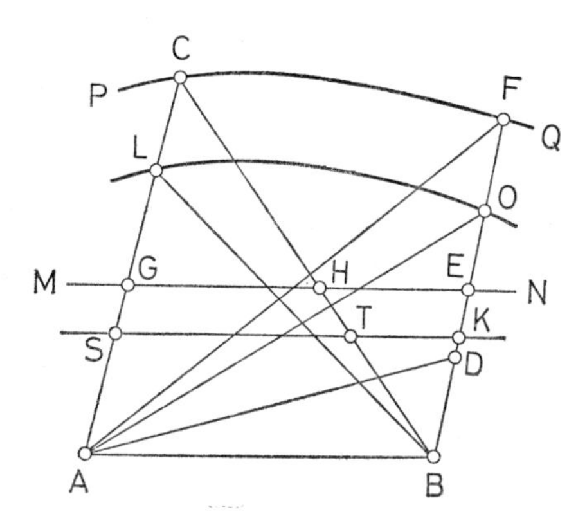

Figure 30

For let MN bisect $\overline{AC}$ as well as $\overline{BC}$, and let the curve $\widehat{PQ}$ passing through the point C be parallel to MN. Then D lies on $\widehat{PQ}$.

In fact, if the half-line $\overrightarrow{BD}$ intersects the line MN at the point E and therefore, by §39, it intersects $\widehat{PQ}$ at distance $\overline{EF} = \overline{BE}$, then

$$\triangle ABC = \triangle ABF$$

and consequently

$$\triangle ABD = \triangle ABF;$$

thus D coincides with F.

On the other hand, if $\overrightarrow{BD}$ does not intersect MN, let C be the point where the perpendicular bisector of $\overline{AB}$ intersects $\widehat{PQ}$. Choose $\overline{GS} = \overline{HT}$ so that the line ST intersects the produced line BD at a point K (the possibility of this choice follows as in §4). Moreover, let

$$\overline{SL} = \overline{SA}, \quad \widehat{LO} \parallel ST,$$

and let O be the intersection of the straight line BK and the curve $\widehat{LO}$. Then by §39

$$\triangle ABL = \triangle ABO$$

and therefore

$$\triangle ABC > \triangle ABD,$$

which contradicts the hypothesis.

§41

The angle sums of two triangles ABC and DEF of equal area are equal.

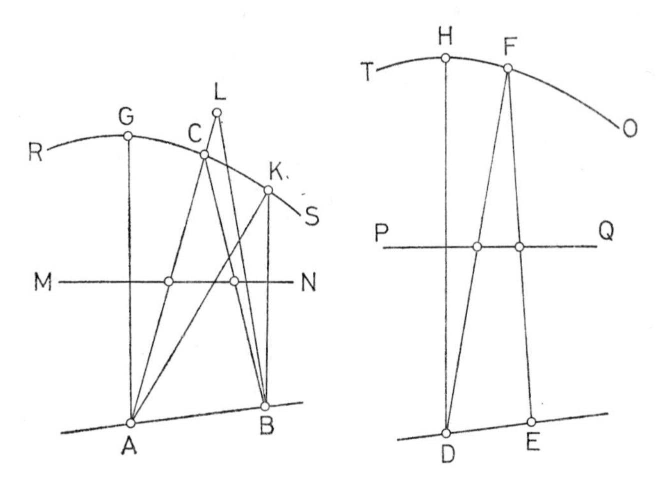

Figure 31

For let MN bisect both $\overline{AC}$ and $\overline{BC}$, and PQ bisect both $\overline{DF}$ and $\overline{EF}$. Further let

$$\widehat{RS} \parallel MN \quad \text{and} \quad \widehat{TO} \parallel PQ.$$

The distance $\overline{AG}$, perpendicular to $\widehat{RS}$, is either equal to the distance $\overline{DH}$, perpendicular to $\widehat{TO}$, or one of them, say $\overline{DH}$, is greater. In each case, the circle with centre A and radius $\overline{DF}$ has a point K in common with $\widehat{GS}$. Then by **§39**

$$\triangle ABK = \triangle ABC = \triangle DEF.$$

But the angle sum of $\triangle AKB$ is by **§40** equal to that of $\triangle DFE$, and by **§39** equal to that of $\triangle ABC$. Hence also $\triangle ABC$ and $\triangle DEF$ have equal angle sums.

In system **S**, this theorem may be reversed. For let the angle sums of $\triangle ABC$ and $\triangle DEF$ be equal and

$$\triangle BAL = \triangle DEF.$$

By the foregoing, the angle sum of one of the latter triangles is equal to that of the other. Consequently, also the angle sum of $\triangle ABC$ is equal to that of $\triangle ABL$. Hence obviously

$$\sphericalangle BCL + \sphericalangle BLC + \sphericalangle CBL = 2R.$$

But according to **§31** the angle sum, in system **S**, of any triangle is less than $2R$. Thus L coincides with C.

§42

If u and v are the supplements to $2R$ of the angle sums in $\triangle ABC$ and $\triangle DEF$, respectively, then

$$\triangle ABC : \triangle DEF = u : v.$$

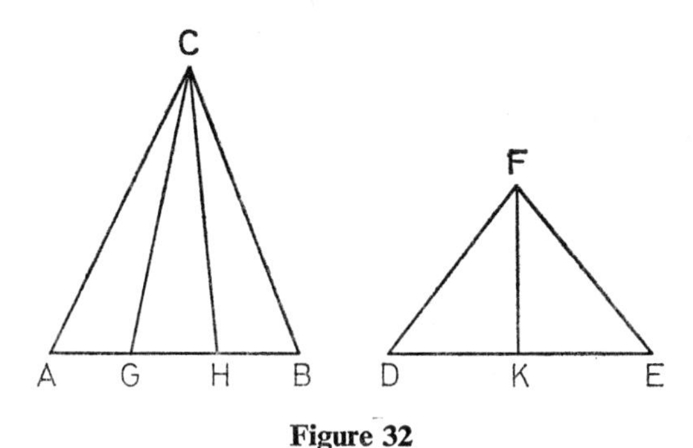

Figure 32

For let each of the triangles ACG, GCH, HCB, DFK, KFE have area p and

$$\triangle ABC = mp, \quad \triangle DEF = np.$$

Further let the angle sum of any triangle of area p be equal to s. Then manifestly

$$2R - u = ms - (m-1)2R = 2R - m(2R - s)$$

i.e.

$$u = m(2R - s),$$

and similarly

$$v = n(2R - s).$$

Thus

$$\triangle ABC : \triangle DEF = m : n = u : v.$$

It is easy to see that this extends to the case of triangles ABC and DEF whose areas are incommensurable.

It can be proved similarly that the areas of two spherical triangles are to each other as are the excesses over $2R$ of their angle sums. If two angles of a spherical triangle are right angles, then the third angle z is just the excess mentioned above. But the area of this spherical triangle is by **§32, (VI)** equal to

$$\frac{z}{2\pi} \frac{p^2}{2\pi},$$

where p denotes the circumference of the great circle. Consequently, the area of any spherical triangle with excess z is

$$\frac{zp^2}{4\pi^2}.$$

§43

We now express the area, in system S, of a rectilinear triangle in terms of the angle sum.

If $\overline{AB}$ increases indefinitely*, then by **§42** the ratio

$$\triangle ABC:(R-u-v)$$

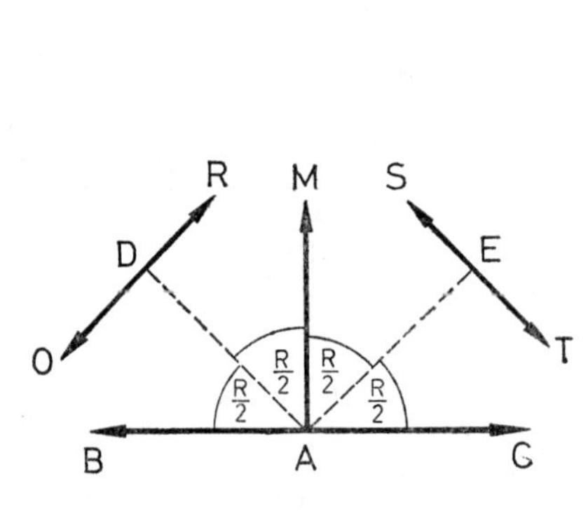

Figure 33

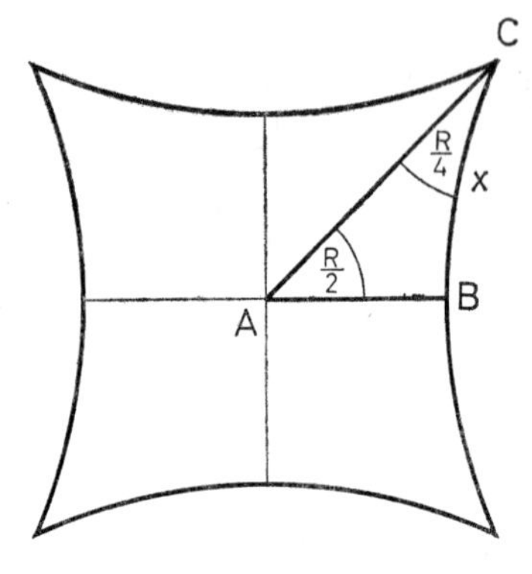

Figure 34

remains constant. On the other hand, by **§32**, (V)

$$\triangle ABC \to BACN$$

and by **§1**

$$R-u-v \to z.$$

Hence

$$BACN:z = \triangle ABC:(R-u-v) = BAC'N':z'.$$

Furthermore, **§30** obviously yields

$$BDCN:BD'C'N' = r:r' = \operatorname{tg} z:\operatorname{tg} z'.$$

But for $y' \to 0$

$$\frac{BD'C'N'}{BAC'N'} \to 1$$

and also

$$\frac{\operatorname{tg} z'}{z'} \to 1.$$

It follows that

$$BDCN:BACN = \operatorname{tg} z:z.$$

By **§32**, however,

$$BDCN = rk = k^2 \operatorname{tg} z.$$

Therefore

$$BACN = zk^2.$$

* See Fig. 20.

Thus, denoting from now onwards the area of any triangle the supplement to $2R$ of whose angle sum is z simply by $\triangle$, we have

$$\triangle = zk^2.$$

Hence it easily follows that if*

$$\overrightarrow{OR} \parallel \overrightarrow{AM} \quad \text{and} \quad \overrightarrow{RO} \parallel \overrightarrow{AB}$$

then the area of the domain enclosed by the lines OR, ST and BC, which is obviously the limit of the areas of indefinitely increasing rectilinear triangles i.e. the limit of $\triangle$ for $z \to 2R$, will be

$$\pi k^2 = \odot\, k$$

for a circular domain of the surface **F**. Denoting this limit by $\square$, from **§30** and **§21** we obtain

$$\pi r^2 = \mathrm{tg}^2\, z \cdot \square = \odot\, r$$

for a circular domain of **F**. By **§32**, (VI) the latter value is equal to $\odot\, s$, where s denotes the chord** $\overline{DC}$. Now if, perpendicularly bisecting the given radius s of the circle in the plane (or the **L**-formed radius of the circle in the surface **F**) and leaning on **§34**, we construct a half-line $\overrightarrow{DB}$ with

$$\overrightarrow{DB} \parallel \simeq \overrightarrow{CN},$$

then by drawing the perpendicular CA to DB and the perpendicular CM to CA we obtain z. Hence, taking any **L**-formed radius for unit, tg z can (by **§37**) be determined geometrically, with the help of two uniform arcs of the same curvature. If we know only their endpoints and construct their axes, such arcs can obviously be compared in measure just as straight line segments, and in this respect they may be considered equivalent to straight line segments.

Now a quadrilateral, say a square, of area $\overline{\square}$ can be constructed as follows. Let***

$$\sphericalangle ABC = R, \quad \sphericalangle BAC = \frac{1}{2} R, \quad \sphericalangle ACB = \frac{1}{4} R,$$

and

$$\overline{BC} = x.$$

By **§31**, (II) X can be expressed using merely square roots; by **§37** it can be constructed as well. If we know X, we can determine x with the help of **§38** (or **§29** and **§35**). Further it is clear that eight times the area of $\triangle\, ABC$ is equal to $\square$.

Thus the geometrical quadrature of the plane circle of radius s is accomplished in terms of a rectilinear figure and uniform arcs of one and the same kind (the latter are equivalent, as concerns comparison, to segments); the complanation of a circular do-

* See Fig. 33.
** See Fig. 20.
*** See Fig. 34.

main of **F** *can be performed likewise. Consequently, either Axiom XI of* EUCLID *holds or the geometrical quadrature of the circle is possible;* though until now it has remained undecided which of these two cases takes place in reality.

If $\mathrm{tg}^2 z$ is an integer or a rational fraction whose denominator (after reducting the fraction to the simplest form) is either a prime number of the form 2^m+1 (of which form is also $2=2^0+1$) or a product of any number of primes of this form where each prime, excepting 2 which alone may occur any number of times, appears as a factor only once, then by the theory of polygons due to the celebrated GAUSS (an outstanding discovery of our age and actually of all times) for these and only these values of z even a rectilinear figure of area $\mathrm{tg}^2 z \cdot \square = \odot s$ can be constructed.

In fact, since the theorem of §**42** can be easily extended to arbitrary polygons, division of the area $\square$ obviously requires dissection of $2R$. However, it is possible to show that this can be achieved in a geometrical way only under the condition stated. On the other hand, in all such cases the foregoing helps to reach the purpose easily. Moreover, if n belongs to the Gauss class, then any rectilinear figure can be converted geometrically into an n-sided regular polygon of equal area.

To settle the matter in all respects, it has remained to prove the impossibility of deciding without any assumption whether Σ or some (and which) **S** is valid. Nevertheless, we leave this to a more appropriate occasion.

PART III

REMARKS

THE HILBERTIAN SYSTEM OF AXIOMS FOR EUCLIDEAN GEOMETRY

FARKAS as well as JÁNOS BOLYAI and, independently, LOBACHEVSKY set themselves the task to give an axiomatic foundation of geometry. In Part I we sketched the history of the evolution; from there it turns out why the realisation of the grand program had been preceded by attempts to clarify the problem of parallel lines. It is worth once more recalling the famous letter of GAUSS, notably the paragraph that begins as follows: "For treating geometry correctly from the outset, it is indispensable ..." (cf. Part I, quotation from the letter of 6th March 1832, last paragraph but one). Nevertheless, the axiomatic foundation of Euclidean geometry was only laid more than half a century later, in 1899, in HILBERT's book *Grundlagen der Geometrie*. HILBERT could already rely on fruitful preparatory work of others. It is sufficient if we refer to the geometrical investigations of RIEMANN, BELTRAMI, LIE, CHASLES, LAGUERRE, CAYLEY, KLEIN, POINCARÉ, CLIFFORD, PASCH, and VERONESE.

In HILBERT's book mentioned above, one finds the first irreproachable system of axioms for Euclidean geometry. When selecting his system of axioms, HILBERT carried out the following program.

First, he defined axiomatically the concepts of point, straight line, plane and space in connection with the relation, called *incidence,* available between them. This was accomplished by a group of requirements consisting of *eight* axioms, the *axioms of incidence*.

On the basis of the axioms of incidence, one may speak of three points of a line, and it becomes possible to introduce a new relation among them: one of the points is *between* the other two. Then one adds to the foregoing group *four* new axioms, the *axioms of order*. The following concepts can be based upon the axioms of order: segment, triangle, half-line, half-plane, half-space, angle and angular domain, polygon and polygonal domain, boundary point, interior point, exterior point, closed region, open region, convexity.

At this stage it is possible to introduce the concepts of the *congruence* (equality) *of segments* and the *congruence* (equality) *of angles*. They can be defined axiomatically; for this purpose, HILBERT added to the preceding twelve axioms *five* more, the *axioms of congruence*. With the help of the 17 axioms obtained in this way, one can define the concepts of smaller and greater for segments or angles and establish, to a certain extent, the segment and angle calculi. Moreover, one can deduce some simple but important theorems, such as the following two:

An exterior angle of a triangle is greater than either of the angles of the triangle that are not adjacent to it.

If each of two intersecting planes is perpendicular to a third plane, then their line of intersection is also perpendicular to that plane.

HILBERT's next axiom defines the concept of *parallelism*.

We replace the last two axioms, the *axioms of continuity*, by an axiom belonging to DEDEKIND. The axiom of continuity makes possible the *measurement* of segments and angles which provides the foundation of *coordinate geometry*.

Omitting the axiom of parallelism from the 19 axioms enumerated so far, theorems deducible from the collection of the remaining 18 axioms are those called *absolute* theorems by BOLYAI. *Hilbert's* system of axioms defines Euclidean geometry in an axiomatic way; the axiom of parallelism makes this system complete.

We divide the axioms of the system into the groups indicated above, and give them numbers.

I. THE AXIOMS OF INCIDENCE

I.1. For two points A and B there is a line a which is incident with both A and B.

I.2. For two points A and B there is no more than one line that is incident with both A and B.

I.3. For any line, there are at least two points incident with it. There are three points not incident with a line.

I.4. For any three points A, B and C not incident with a line there is at least one plane σ incident with A, B and C. For any plane there is at least one point incident with it.

I.5. For any three points A, B and C not incident with a line there is no more than one plane incident with A, B and C.

I.6. If two points A and B of the line a are incident with the plane σ, then all points of a are incident with σ.

I.7. If two planes α and β have a point P in common, they have at least one more point Q in common.

I.8. There are at least four points not incident with a plane.

II. THE AXIOMS OF ORDER

II.1. If the point B lies between the points A and C, then A, B and C are three distinct points of a line and, moreover, B lies between C and A.

II.2. For two points A and C there is at least one point B on the line AC which lies between A and C.

II.3. Of any three points on a line, one and only one lies between the other two.

II.4. If the points A, B and C are not incident with a line, and a is a line in the plane ABC not incident with A, B and C but incident with a point between A and B, then a is also incident with either a point between B and C or a point between C and A.

III. THE AXIOMS OF CONGRUENCE

III.1. If AB is a segment and a' is a half-line with endpoint A', then there is a point B' on a' such that the segment AB is congruent (equal) to the segment $A'B'$. We write

$$AB \equiv A'B'.$$

III.2. If the segments $A'B'$ and $A''B''$ are congruent to one and the same segment AB, then the segment $A'B'$ is congruent to the segment $A''B''$.

III.3. If the segments AB and BC of the line a have no interior point in common and the segments $A'B'$ and $B'C'$ of the same or another line a' have no interior point in common, while $AB \equiv A'B'$ and $BC \equiv B'C'$, then $AC \equiv A'C'$.

III.4. Let (h, k) be a convex angle. Consider a half-plane α', a point O' on the boundary of α', and a half-line h' that belongs to the boundary of α' and has endpoint O'. In the half-plane α' there is one and only one half-line k' having endpoint O' and such that the angle (h, k) is congruent (equal) to the angle (h', k'). We write

$$\sphericalangle(h, k) \equiv \sphericalangle(h', k').$$

III.5. If for two triangles ABC and $A'B'C'$ we have the congruences

$$AB \equiv A'B', \quad AC \equiv A'C' \quad \text{and} \quad \sphericalangle BAC \equiv \sphericalangle B'A'C',$$

then also

$$\sphericalangle ABC \equiv \sphericalangle A'B'C'$$

is valid.

IV. THE AXIOM OF PARALLELISM

IV. In the plane determined by the line a and the point A not incident with a, among all lines incident with A there is at most one which does not intersect a.

V. THE AXIOMS OF CONTINUITY

V.1. Given any two segments AB and CD, there are distinct points $A_1, A_2, \ldots, A_n$ on the line AB such that each of the segments $AA_1, A_1A_2, A_2A_3, \ldots, A_{n-1}A_n$ is congruent to the segment CD, and B lies between A and A_n.

V.2. Adding new points or new lines to the plane, the extended system cannot be a plane for which all the foregoing axioms are valid.

The axioms of continuity may be replaced by Dedekind's axiom which reads as follows.

V. If the points of the line are divided into two classes such that neither class is empty and neither class contains a point between two points of the other class, then there is a point that lies between any two points different from it and belonging to different classes.

REMARKS TO §§ 1—43

By stating the system of axioms of Euclidean geometry, especially the axioms of the groups I, II, III and V which provide the basis of absolute geometry, we have intended to give an idea of the foundations necessary for an up-to-date treatment of Bolyai's work. However, a systematic account of the most significant theorems deducible from the axioms is beyond the scope of this work: it would require a separate volume. For such an account we refer to the book *Les fondements de la géometrie,* I—II (Budapest, 1955–1966, Akadémiai Kiadó) by BÉLA KERÉKJÁRTÓ.

Next, following the order of BOLYAI's work, we set forth our explanatory, supplementary, orientative remarks.

§1

In this section, BOLYAI defines parallelism in a way compatible with the residual system of axioms. His definition of parallelism is wider than the Euclidean one.

§2

In the definition stated in the previous section, the point B has a distinguished role as the endpoint of the half-line $\overrightarrow{BN}$. In this section it is proved that *parallelism is independent of the choice of B on the line containing the half-line* $\overrightarrow{BN}$.

§3

In Euclidean geometry there is a theorem which says that two lines parallel to a third one are parallel to each other. In this section, for the new definition of parallelism, it is only proved that *two half-lines parallel to a third one do not intersect each other.*

It is even now true that two half-lines of the latter kind, and the lines containing them, are not only non-intersecting but also parallel. This, however, is proved later, in §7.

§4

This section deals with proving the theorem that *if* (see Fig. 3 to Part II) *the angle* MAN *is shifted along its arm* $\overrightarrow{AM}$ *then its other arm* $\overrightarrow{AN}$ *sweeps out the angular domain* MAN *in a one-to-one manner, without gaps.* Strictly speaking, only the absence of gaps is proved. That the sweeping is one-to-one can also be seen immediately. Otherwise, in fact, there is a point B in the angular domain to which there correspond two points C and C' on $\overrightarrow{AM}$ so that

$$\sphericalangle BCM = \sphericalangle NAM = \sphericalangle BC'M.$$

Since

$$\sphericalangle CC'B + \sphericalangle BC'M = 2R,$$

it follows that

$$\sphericalangle CC'B + \sphericalangle BCM = 2R.$$

Thus the sum of two angles of the triangle BCC' is $2R$, which is impossible by the residual system of axioms.

§5

This section discusses, with the help of a continuity theorem, the existence of an isogonal correspondence between the points of two parallel lines.

The theorem in question can be deduced also without reference to continuity, relying only on the axioms of congruence. We prove the theorem in this way and verify even the uniqueness of the correspondence. We make use of the fact, obtainable through the axioms of congruence from the residual system of axioms alone, that the points which are at equal distance from the arms of an angle lie on the bisector of the angle and each point of the bisector has this property.

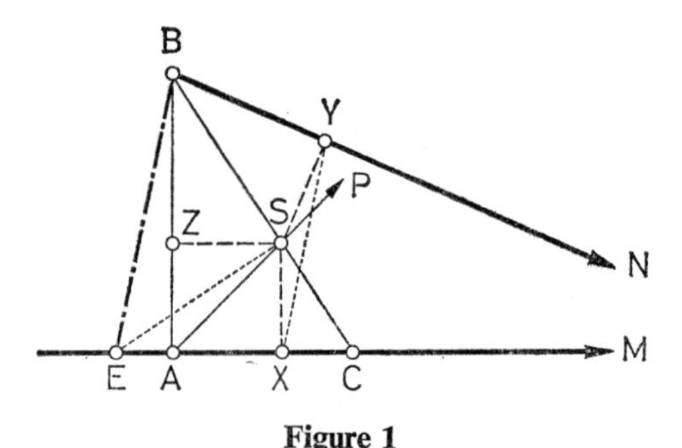

Figure 1

Since $\overrightarrow{BN} \| \overrightarrow{AM}$, the bisector of $\sphericalangle ABN$ intersects $\overrightarrow{AM}$ at a point C. The bisector $\overrightarrow{AP}$ of the angle BAC of the triangle ABC intersects the segment BC at a point S. Let.

$$SX \perp AM, \quad SY \perp BN, \quad SZ \perp AB.$$

Then

$$SX = SZ = SY.$$

Thus $SX = SY$.

Let E be the point on $\overrightarrow{XA}$ with

$$YB = XE.$$

Obviously

$$\triangle SXE \equiv \triangle SYB;$$

hence $\triangle ESB$ is isosceles. From these three triangles it is clear that

$$\sphericalangle BEM = \sphericalangle EBN.$$

As a result, on AM an isogonal companion of B can be found (GAUSS terms it *corresponding point*).

Moreover, there is only one such companion point. For, if F and G are two points corresponding to B on the line AM and G belongs to $\overrightarrow{FM}$, then

$$\sphericalangle BFM = \sphericalangle FBN, \quad \sphericalangle BGM = \sphericalangle GBN,$$

$$\text{and} \quad \sphericalangle FBN > \sphericalangle GBN.$$

Consequently,

$$\sphericalangle BFM > \sphericalangle BGM = 2R - \sphericalangle BGF$$

and therefore $\sphericalangle BFM + \sphericalangle BGF > 2R$, which cannot hold, as consideration of the triangle BFG shows.

§6

In this section it is shown that *parallelism is independent of the endpoints of the half-lines*. Further, we learn that *parallelism is a symmetric relation*. The deduction of the latter circumstance is based on the concept of corresponding points. Here we give a proof which does not use this concept.

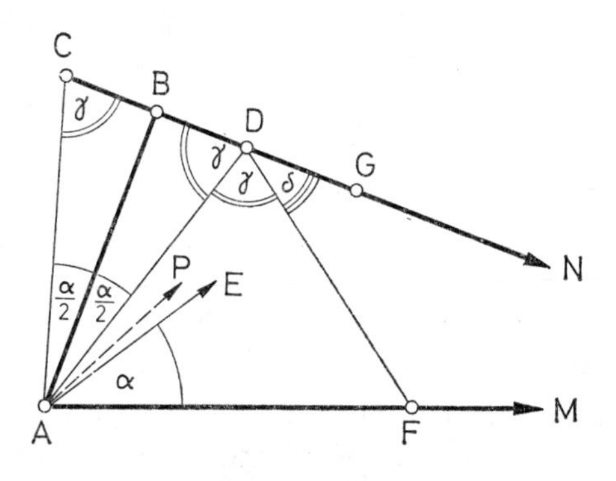

Figure 2

Let $\overrightarrow{BN} \parallel \overrightarrow{AM}$. Assume that $\overrightarrow{AM}$ is not parallel to $\overrightarrow{BN}$. Since they cannot have any point in common, they are non-intersecting. Then the half-line $\overrightarrow{AP}$ satisfying $\overrightarrow{AP} \parallel \overrightarrow{BN}$ lies in $\sphericalangle BAM$.

Let C and D be points on the line BN sufficiently close to B and such that

$$\sphericalangle CAB = \sphericalangle DAB \quad \text{and} \quad \sphericalangle CAD < \sphericalangle PAM.$$

We construct the half-line $\overrightarrow{AE}$ for which

$$\sphericalangle EAM = \sphericalangle CAD;$$

then $\overrightarrow{AE}$ lies in the interior of $\sphericalangle PAM$.

Let F on $\overrightarrow{AM}$ satisfy

$$\sphericalangle BDA = \sphericalangle ADF$$

(this can be achieved since the sum of two angles of $\wedge ACD$ is $< 2R$ and $\overrightarrow{DN} \parallel \overrightarrow{AM}$). Further let

$$CG = DF.$$

Then

$$\triangle ACG \equiv \triangle ADF,$$

which implies

$$\sphericalangle CAG = \sphericalangle DAF$$

and therefore

$$\sphericalangle GAM = \sphericalangle CAD = \sphericalangle EAM.$$

Thus $\overrightarrow{AE}$ passes through G and consequently $\overrightarrow{AP}$, which lies in $\sphericalangle BAE$, also intersects $\overrightarrow{CN}$, contrary to the assumption.

§7

In this section it is proved that *parallelism is a transitive relation.*

The proof first deals with the case where the three lines are in one and the same plane, and then with the general case. The first case is divided into two subcases: two lines either lie on different sides or on the same side of the third line parallel to each of them.

Bolyai himself points out (in the Errata) that the subdivision of the first case can be avoided if the proof is first carried out for three non-coplanar lines. The proof for the case of coplanar lines can be based, without subdivision, on the theorem proved for the non-coplanar case.

This is the first occurrence of the peculiar method that consists in deducing planimetric relations from stereometric ones (naturally, the proof of the relevant stereometric relations is first needed). This method is often applied in the sequel.

We demonstrate the transitivity of parallelism for coplanar lines assuming that it is already proved for the non-coplanar case.

Let $\overrightarrow{AM}$, $\overrightarrow{BN}$ and $\overrightarrow{CP}$ lie in the plane σ and let D be a point outside σ. If

$$\overrightarrow{BN} \parallel \overrightarrow{AM}, \quad \overrightarrow{CP} \parallel \overrightarrow{AM} \quad \text{and} \quad \overrightarrow{DQ} \parallel \overrightarrow{AM},$$

then applying the theorem already proved to the non-coplanar triples $\overrightarrow{AM}$, $\overrightarrow{BN}$, $\overrightarrow{DQ}$;

$$\overrightarrow{AM}, \overrightarrow{CP}, \overrightarrow{DQ} \quad \text{and} \quad \overrightarrow{BN}, \overrightarrow{CP}, \overrightarrow{DQ},$$

respectively, we obtain that

$$\overrightarrow{BN} \parallel \overrightarrow{DQ}, \quad \overrightarrow{CP} \parallel \overrightarrow{DQ} \quad \text{and} \quad \overrightarrow{BN} \parallel \overrightarrow{CP}.$$

§8

In this section it is shown that, *for an isogonal secant of two parallel lines, the perpendicular bisector of the segment of the secant between the two parallel lines is parallel to both of these lines.*

The proof exploits the fact that the broken line formed by $\overrightarrow{BN}$, $\overrightarrow{CP}$ and $\overline{BC}$ is symmetric with respect to the perpendicular bisector of $\overline{BC}$. This symmetry yields also the following theorems immediately:

A. *The mid-point of the segment of any isogonal secant of $\overrightarrow{BN}$ and $\overrightarrow{CP}$ bounded by these two parallel lines belongs to the perpendicular bisector of $\overline{BC}$.*

B. *If a line which is perpendicular to the perpendicular bisector of $\overline{BC}$ intersects one of the lines BN and CP, then it also intersects the other; moreover, it forms equal angles with both lines.*

C. *The segments of two parallel lines bounded by two isogonal secants are equal.*

We shall call the locus of the mid-points of all segments contained in the isogonal secants of two parallel lines and bounded by these lines the *mid-line of the strip formed by the two parallel lines.*

§9

According to the theorem proved in this section, *if the interior dihedral angles between the plane of two parallel lines and two planes incident with the two lines, respectively, have sum $<2R$, then the two latter planes intersect each other.* Bolyai restricts his attention to the special case where one of the dihedral angles considered is $=R$. He leaves the general case to the reader. Let us see the proof in this case.

If $\overrightarrow{AM} \parallel \overrightarrow{BN}$, and the plane β is incident with $\overrightarrow{BN}$, then there is one and only one plane α which is incident with $\overrightarrow{AM}$ and does not intersect β.

For AMC, the plane of projection of $\overrightarrow{AM}$, is perpendicular to β and intersects β at the projection $\overrightarrow{CP}$ of $\overrightarrow{AM}$. Further $\overrightarrow{CP}$ is parallel to both $\overrightarrow{AM}$ and $\overrightarrow{BN}$ (this follows

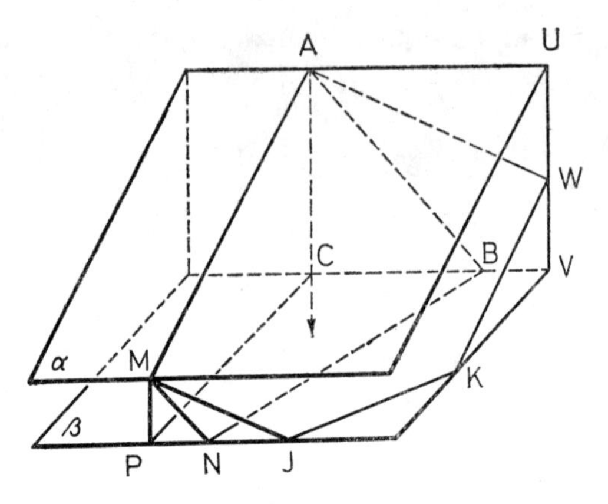

Figure 3

from §7, making use of the assumption $\overrightarrow{AM} \| \overrightarrow{BN}$). The plane of the strip (AM, CP) is perpendicular to β and therefore, by the special theorem proved in §9, a plane α that is incident with the boundary half-line $\overrightarrow{AM}$ of this strip will not intersect β if and only if $\alpha \perp AMC$.

Assume that $\overrightarrow{AM} \| \overrightarrow{BN}$, *and let the plane* β *be incident with* $\overrightarrow{BN}$. *If* τ *denotes the reflection of* β *in the mid-line of the strip* (AM, BN), *then* τ *is incident with* $\overrightarrow{AM}$ *and does not intersect* β.

The first half of the proposition is obvious. The second half needs a proof only if β does not coincide with the plane of the strip (AM, BN). In the latter case, suppose τ intersects β in a line t. The reflection in the mid-line of the strip transforms β and τ into τ and β, respectively, thus it transforms t into itself. Reflection in a line leaves invariant only the lines that intersect the line of reflection perpendicularly; it also leaves invariant the point at which the line of reflection and a perpendicular line intersect each other. Consequently, t intersects the mid-line of the strip perpendicularly at some point T. The point T belongs to each of β, τ and the plane of the strip. But the existence of a point of this kind contradicts the hypothesis $\overrightarrow{AM} \| \overrightarrow{BN}$. Thus the assumption that τ intersects β has been false.

Comparing these two theorems we obtain:

If $\overrightarrow{AM} \| \overrightarrow{BN}$, *and* α *incident with* $\overrightarrow{AM}$ *does not intersect* β *incident with* $\overrightarrow{BN}$, *then a half-turn about the mid-line of the strip* (AM, BN) *carries* β *into* α.

Clearly, the planes α and β form such dihedral angles with the plane of the strip that their sums, on each side of the latter plane, are equal to $2R$. If, say, α turns about AM starting from the unique position in which it does not intersect β, then one of the sums becomes smaller. This completes the proof in the general case.

This section discusses the proof of the theorem which says that *correspondence* (between the isogonal points of parallel lines) *is a transitive relation.*

If the spatial case is proved first, then it can be used when deriving the theorem in the planar case. Nevertheless, one may ask how to obtain the theorem for coplanar triples independently of these spatial relations. Before treating this question, we prove some absolute theorems of plane geometry.

A. *If the bisectors of two exterior angles of a triangle are parallel to each other, then the bisector of the interior angle at the third vertex is also parallel to them.*

For let ABC be the triangle, and let AM and BN denote the bisectors of its exterior angles at A and B, respectively. If the half-lines $\overrightarrow{AM} \| \overrightarrow{BN}$ are on the side opposite to C of the line AB, then the bisector of $\sphericalangle BCA$ enters $MABN$ at some point D which lies between A and B.

Let P be a point in $MABN$ such that

$$\overrightarrow{AM} \| \overrightarrow{DP} \| \overrightarrow{BN}.$$

If $\overrightarrow{CD}$ does not contain $\overrightarrow{DP}$, then it enters either $MADP$ or $PDBN$, and by **§1** intersects either $\overrightarrow{AM}$ or $\overrightarrow{BN}$ at a point.

If K is a common point of $\overrightarrow{CD}$ and $\overrightarrow{AM}$, then the distances from K to the three side lines of the triangle are equal. In fact, the well-known theorem which states that the points of the bisector of an angle are at equal distances from the arms does not rely on the axiom of parallelism. Hence K is on $\overrightarrow{BN}$, contrary to the assumption. Similarly, the assumption that $\overrightarrow{CD}$ intersects $\overrightarrow{BN}$ leads to a contradiction.

Thus $\overrightarrow{CD}$ contains $\overrightarrow{DP}$.

B. *If there exist two parallel lines which are intersected by one of their isogonal secants at right angles, then any pair of parallel lines is intersected by any isogonal secant at right angles.*

For let

$$\overrightarrow{AM} \| \overrightarrow{CP} \quad \text{and} \quad \sphericalangle CAM = \sphericalangle ACP < R.$$

Then, obviously, the perpendicular bisector $\overrightarrow{DS}$ of AC and the half-line $\overrightarrow{BN}$ that is perpendicular to AC at B (where $AC = CB$) are also parallel to $\overrightarrow{CP}$ (§8). Hence it follows at once that the perpendiculars to the line AC through the members of a point sequence on AC obtained from the segment $\overline{AC}$ with the help of reflections and bisections are parallel to one another. Thus we have a pencil, however dense, of parallel lines intersected by one and the same line at right angles.

Further let

$$\overrightarrow{A_1M_1} \| \overrightarrow{B_1N_1} \quad \text{and} \quad \sphericalangle B_1A_1M_1 = \sphericalangle A_1B_1N_1 < R.$$

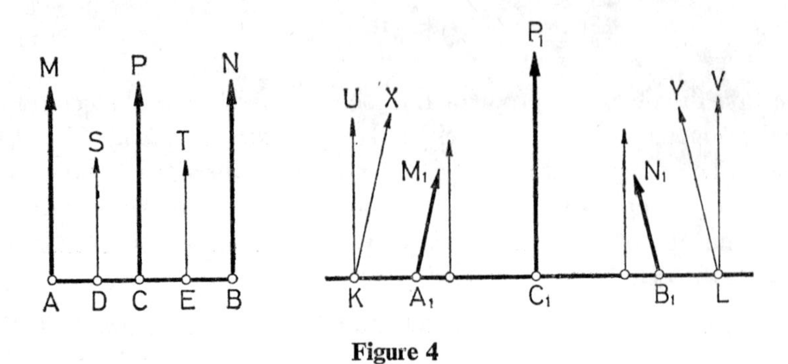

Figure 4

Let, moreover, C_1P_1 be the mid-line of the strip (A_1M_1, B_1N_1) Consider the segment

$$\frac{m}{2^n} \cdot AC = C_1K > C_1A_1,$$

where $0 < m, n$ are integers. By the preceding paragraph, if $\overrightarrow{KU}$ is perpendicular to KC_1 then it is parallel to $\overrightarrow{C_1P_1}$. Now if

$$\sphericalangle C_1A_1M_1 = \sphericalangle C_1KX,$$

then $\overrightarrow{KX}$ lies in C_1KU and therefore intersects $\overrightarrow{C_1P_1}$ at some point Q. The half-line $\overrightarrow{A_1M_1}$ enters the triangle KQC_1, hence also leaves it somewhere; as it does not intersect $\overrightarrow{C_1Q}$, it ought to intersect KX at a point O. This, however, is impossible, since the sum of the angles at K and C_1 of the triangle OKC_1 cannot be equal to $2R$.

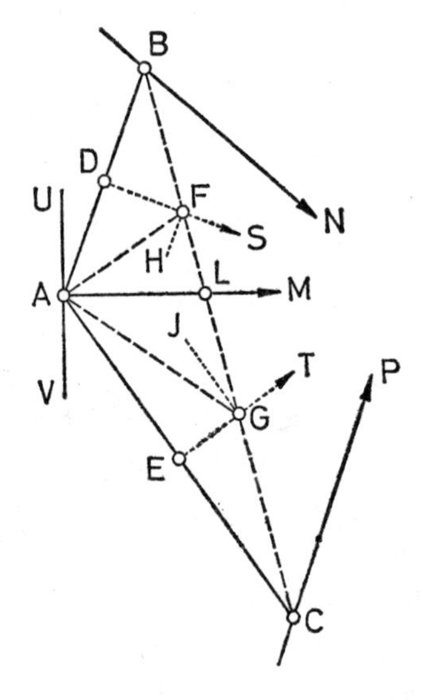

Figure 5

Thus $\sphericalangle C_1 A_1 M_1$ can only be a right angle.

We finally return to proving the transitivity of the relation $\| \simeq$ for coplanar lines.

Let the half-lines $\overrightarrow{AM}, \overrightarrow{BN}$ and $\overrightarrow{CP}$ be parallel to one another. In addition, suppose that

$$\overrightarrow{BN} \simeq \overrightarrow{AM}, \quad \overrightarrow{CP} \simeq \overrightarrow{AM}.$$

Case a). Let $\overrightarrow{AM}$ lie in the strip (BN, CP). If UV passes through A and is perpendicular to $\overrightarrow{AM}$, while $\overrightarrow{AU}$ contains AB, then by Theorem **B** also $\overrightarrow{AV}$ contains AC, so that the transitivity theorem is clear. On the other hand, if AB lies in $\sphericalangle UAM$ then, again by Theorem **B**, AC lies in $\sphericalangle VAM$. In this case, AM separates B and C; hence $\overrightarrow{AM}$ intersects BC at some point L between B and C.

The mid-line $\overrightarrow{DS}$ of (AM, BN) leaves the triangle ABL somewhere. Being parallel to $\overrightarrow{AM}$, it can leave ABL only across the side BL, at a point F between B and L. Similarly, the mid-line of (AM, CP) intersects CL at some G lying between C and L.

By reasons of symmetry,

$$\sphericalangle CBN = \sphericalangle FAL \quad \text{and} \quad \sphericalangle BCP = \sphericalangle GAL.$$

The bisectors of the angles at F and G of the triangle AFG are $\overrightarrow{FS}$ and $\overrightarrow{GT}$. According to §§7 and **8**

$$FS \| GT.$$

Therefore, by Theorem **A**, the half-line $\overrightarrow{AL}$ which is parallel to them is an angle-bisector of FAG. Hence

$$\sphericalangle FAL = \sphericalangle GAL,$$

which yields

$$\sphericalangle CBN = \sphericalangle BCP.$$

Thus, in fact, $\overrightarrow{BN} \simeq \overrightarrow{CP}$.

Case b). Let $\overrightarrow{AM}$ lie outside the strip (BN, CP). If, say, $\overrightarrow{CP}$ is between $\overrightarrow{AM}$ and $\overrightarrow{BN}$ then, after interchanging notation for $\overrightarrow{AM}$ and $\overrightarrow{CP}$, the assumption reads

$$\overrightarrow{BN} \simeq \overrightarrow{CP}, \quad \overrightarrow{AM} \simeq \overrightarrow{CP}.$$

Let B^* be a point of the line BN such that

$$\overrightarrow{B^*N} \simeq \overrightarrow{AM}.$$

Then from a), on account of the relations $\overrightarrow{CP} \simeq \overrightarrow{AM}$ and $\overrightarrow{B^*N} \simeq \overrightarrow{AM}$, it follows that not only $\overrightarrow{BN} \simeq \overrightarrow{CP}$ but also $\overrightarrow{B^*N} \simeq \overrightarrow{CP}$. Since the correspondence between isogonal points is one-to-one, B coincides with B^*. Thus

$$\overrightarrow{BN} \simeq \overrightarrow{AM}.$$

§11

In this section the *parasphere* (**F**-surface) and the *paracycle* (**L**-line) are defined. Their definitions are based on the notion of *corresponding points* (see §5).

From the Euclidean point of view, the straight line and the plane, respectively, may be considered as a circle and a sphere of infinite radii. More exactly, this means the following. Let B lie on $\overrightarrow{AM}$ and b be the circle through A with centre B. If B tends to infinity over the points $B, C, D, \ldots$, then b tends, over the circles $b, c, d, \ldots$, to the line v that is perpendicular to $\overrightarrow{AM}$ at the point A (see Fig. 6).

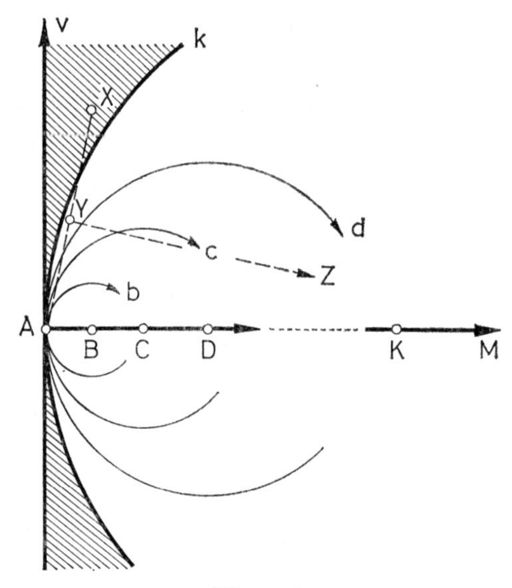

Figure 6

If, for example, we choose a point X in the region between the line v and the circle k, however large the circle may be we can find a circle through X in the pencil considered. Its centre lies where $\overrightarrow{AM}$ and the perpendicular bisector $\overrightarrow{YZ}$ of AX intersect each other. In Euclidean geometry this point of intersection exists as long as X does not belong to v. The views "the circle of infinite radius is a straight line" or "the Euclidean axiom of parallelism holds" are equivalent.

In absolute geometry, the geometry built upon the residual system of axioms, one has to say that the paracycle, the circle of infinite radius, is a straight line or a curve. In hyperbolic geometry it is necessarily a curve; LOBACHEVSKY calls it *horocycle* (limit circle).

Even opposing GAUSS, JÁNOS BOLYAI firmly criticized the term "circle of infinite radius" for the paracycle. He pointed out several times that the concept of circle implies the concept of centre and that of the equality of (finite) distances, and therefore the term in question makes the concept of **L**-line obscure. Consequently, the

L-line should not be introduced as the limit of the pencil of circles, but independently of the circle, namely by the property that the perpendicular bisectors of the chords of this curve are parallel lines. This way of introducing the L-line, in turn, requires previous introduction of the concept of corresponding points as well as a treatment of the most significant facts concerning correspondence.

Prior to WEIERSTRASS, it would have been unusual to insist on such a strict exactitude of mathematical concept-building as BOLYAI deemed necessary. BOLYAI recognized the importance of a rigorous introduction of concepts while he was seeking the way out of the labyrinth of investigations followed by former researchers and, no doubt, at the beginning also by himself.

If we have already built up the concept of paracycle and derived the most important and comprehensive relations for it, then we see to what extent the paracycle and the circle *do or do not resemble* each other. Then in our argument we shall no more carry over the theorems that are characteristic for the Euclidean point of view to the unfamiliar terrain of non-Euclidean geometry, but our intuition will widen and admit a new circle concept which unites the (old) concept of circle with that of the paracycle. We shall know exactly in which problems they should be kept together or separated from each other. By attaching the hypercircle, the concept of circle widens in BOLYAI's treatment still further, becomes saturated — namely the residual axioms do not allow the conceptual existence of any other line that could be shifted within itself — and gives rise to the comprehensive new concept of *uniform* curve.

These were the unusual, strict ideas BOLYAI strove to implement in the *Appendix*. The contents of **§11** are the following.

A. We define the *parasphere* (**F**) belonging to the point A of a given half-line $\overrightarrow{AM}$ as the collection of A and all points (in space) corresponding to A on the lines parallel to $\overrightarrow{AM}$. The half-line $\overrightarrow{AM}$ is the axis of the parasphere.

B. The plane through two axes of the parasphere intersects the parasphere in a curve called *paracycle*.

An independent definition of the paracycle (**L**) is the following: given a half-line $\overrightarrow{AM}$ in the plane, the collection of A and all points corresponding to A on the half-lines parallel to $\overrightarrow{AM}$ in the plane is said to be the paracycle with axis $\overrightarrow{AM}$.

C. The paracycle is divided by any of its axes into two congruent halves.

D. When a paracycle of the parasphere rotates about one of its axes, it sweeps the parasphere.

We show that, in the sense described above, the paracycle may be called "circle of infinite radius"; namely we establish the following theorem.

If $\overrightarrow{BN} \parallel \simeq \overrightarrow{AM}$ and C lies on $\overrightarrow{BN}$, then there is a point O on $\overrightarrow{AM}$ such that $OA=OC$. On the other hand, for a point D on the extension (in the opposite direction) of $\overrightarrow{BN}$ there is no point O with $OA=OD$.

The assertion follows at once from the equivalence of the condition $OA=OC$ to the requirement $\sphericalangle OAC = \sphericalangle OCA$. It is therefore sufficient to show that if we lay

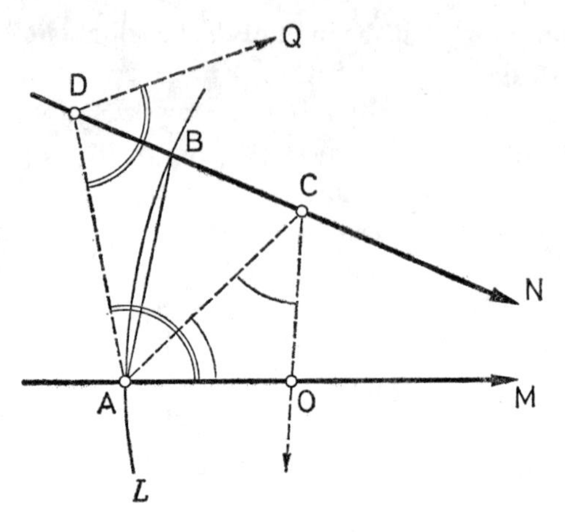

Figure 7

$\sphericalangle MAC$ at C starting from the arm $\overrightarrow{CA}$ in the direction of $\overrightarrow{BN}$ then the arm we obtain does intersect $\overrightarrow{AM}$, while performing a similar construction with D no intersection arises. Indeed, the hypothesis $\overrightarrow{AM} \simeq \overrightarrow{BN}$ yields $\sphericalangle MAC < \sphericalangle ACN$ and $\sphericalangle MAD > > \sphericalangle ADN$. Since $\overrightarrow{AM} \| \overrightarrow{BN}$, the half-line $\overrightarrow{CO}$ which lies between the arms of $\sphericalangle ACN$ does, while $\overrightarrow{DQ}$ which lies outside $\sphericalangle ADN$ does not, intersect $\overrightarrow{AM}$.

§12

If $\overrightarrow{AM} \| \simeq \overrightarrow{BN}$, then the paracycles that belong to $\overrightarrow{AM}$ and $\overrightarrow{BN}$ coincide. Replacing "paracycle" by "parasphere" we obtain the corresponding theorem of stereometry.

Thus, while in the definition of the paracycle or parasphere the axis $\overrightarrow{AM}$ given in advance played a distinguished role — we can say this axis induced the paracycle or the parasphere —, in the present section it becomes clear that from the geometrical point of view the axis inducing the figure has no distinguished role. In other words, distinction was due to the way of formulation just as was the distinguished role of the origins specifying the half-lines in the definition of parallelism (**§1**).

§§ 13-14

These two sections deal with an "inducing" theorem. Contents:

If there exist a pair of parallel lines and a transversal of them such that the sum of the interior angles is equal to 2R, then for any pair of parallel lines and any transversal of them the sum of the interior angles is equal to 2R.

If there exist a pair of parallel lines and a transversal of them such that the sum of the interior angles on one side of the transversal is less than $2R$, *then for any pair of parallel lines and any transversal of them the sum of the interior angles on one side of the transversal is less than* $2R$.

§15

Leaning on the two previous sections, in this section it is declared that one may equally assume either the validity or the falseness of Postulate 5.

(Euclidean) geometry deducible from the first assumption is called *System* Σ, (hyperbolic) geometry deducible from the second is *System* S, while theorems that do not rely on any of these assumptions are called *absolute theorems*.

Evidently, absolute theorems are valid both in Σ and S, that is, *absolute theorems are common theorems of Euclidean and hyperbolic geometries.*

§16

In Euclidean geometry the paracycle to axis $\overrightarrow{AM}$ *is the straight line through* A *perpendicular to* AM.

On the other hand, *in hyperbolic geometry neither the paracycle nor the parasphere have three collinear points.*

§17

In this and the two next sections, relations valid in hyperbolic geometry are only considered. The most important formal properties of the paracycle and the parasphere are set forth.

1°. *In hyperbolic geometry, the paracycle is a curved line and the parasphere is a curved surface.*

2°. *The paracycle is a line which is uniform at all points.*

3°. *The parasphere is a surface which is uniform at all points.*

4°. *The parasphere can always be turned about its suitable axis so that a prescribed point of the parasphere turns into another prescribed point of the parasphere and no point leaves the surface.*

It is now advisable to discuss in some detail what the word *uniform* means. We first consider the structure of the paracycle.

Let A and A' be two points of L, and assume that for the respective axes

$$\overrightarrow{AM} \simeq \| \overrightarrow{A'M'}.$$

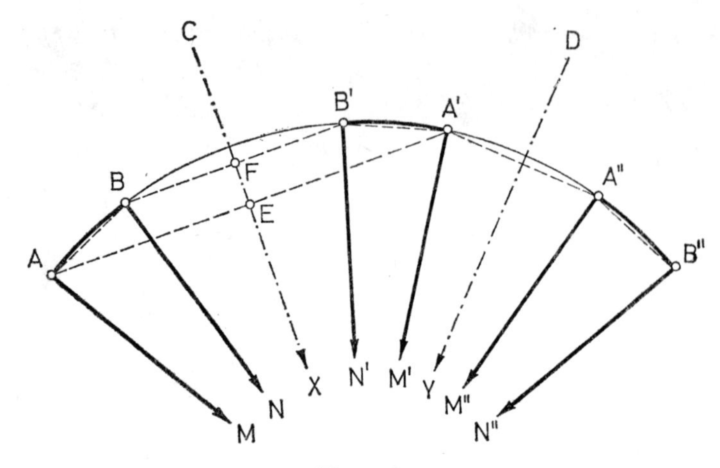

Figure 8

Let $\overrightarrow{CX}$ be the mid-line of the strip formed by these axes. Consider a third point B of **L**. For the corresponding axis

$$\overrightarrow{AM} \parallel \simeq BN.$$

Further, if $\overrightarrow{B'N'}$ is the reflection of $\overrightarrow{BN}$ in $\overrightarrow{CX}$, then

$$\overrightarrow{BN} \parallel \simeq \overrightarrow{B'N'}.$$

With the help of congruent triangles arising by the reflection it is easy to verify that

$$\sphericalangle MAB = \sphericalangle NBA = \sphericalangle M'A'B' = \sphericalangle N'B'A'$$

and

$$AB = A'B'.$$

This reflection maps the paracycle onto itself, takes any chord of the paracycle into a congruent chord, and changes the orientation of the paracycle to the opposite.

The reflection of the paracycle in one of its axes or a half turn about this axis have the same effect (the plane of the paracycle is turned in space).

Let us now reflect A' and B' in another axis $\overrightarrow{DY}$. Then

$$M''A''B''N'' \equiv M'A'B'N' \equiv MABN$$

and the second reflection restores the original orientation of the paracycle. Obviously, the displacements of the initial point and the endpoint are equal:

$$AA'' = BB''.$$

Thus *the paracycle is a line which can be shifted within itself.* (A shift takes any of its chords into a congruent and equally oriented chord.)

We now consider the following configuration in space. Let

$$\overrightarrow{AM} \parallel \simeq \overrightarrow{BN},$$

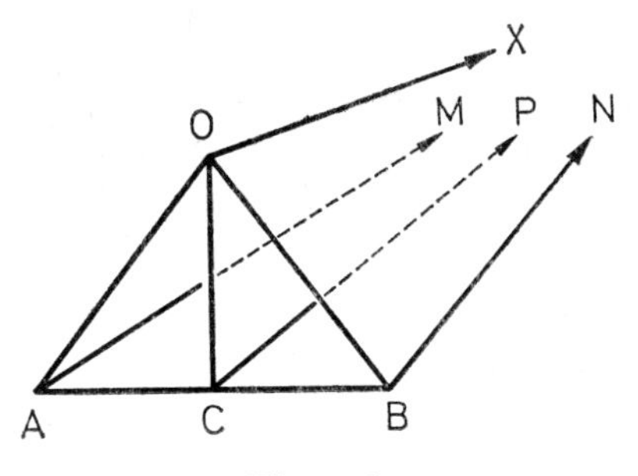

Figure 9

$\overrightarrow{CP}$ the mid-line of the strip (AM, BN), and

$$AC = CB, \quad AB \perp CO \perp CP.$$

Let O be any point in the plane OCP and let

$$\overrightarrow{OX} \parallel \overrightarrow{CP}.$$

With the help of the congruent triangles so arising it is easy to see that

a) planar reflection in (planar symmetry with respect to) OCP takes $MAOX$ into $NBOX$;

b) $MAOX$ can be turned about $\overrightarrow{OX}$ into $NBOX$.

Applying the facts just considered to the paracycle induced by $\overrightarrow{AM}$, we obtain the following theorems.

The parasphere exhibits planar symmetry with respect to each of its principal planes. (A plane is principal if it is incident with an axis of the surface.)

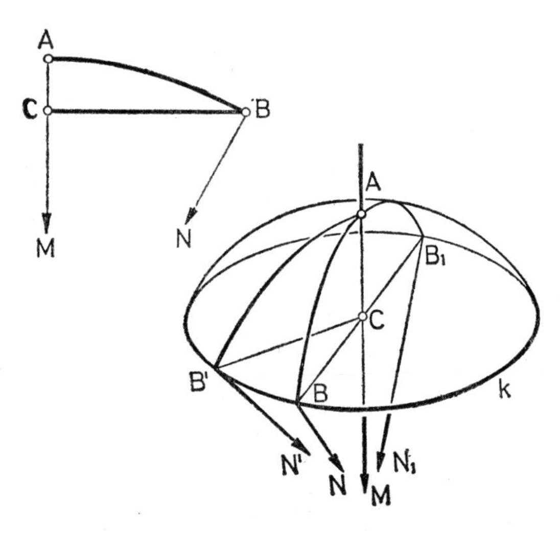

Figure 10

The parasphere can be turned — about any of its axes incident with the perpendicularly bisecting plane of a segment whose endpoints belong to the parasphere — within itself so that one of these endpoints is taken into the other.

Finally, we introduce a configuration which later will play an important role.

Let $\overrightarrow{BN} \| \simeq \overrightarrow{AM}$ and $\sphericalangle MCB = R$. If this plane figure rotates about $\overrightarrow{AM}$, then the paracycle arc $\widehat{AB}$ sweeps a parasphere cap, whereas

the segment $\overline{CB}$ sweeps a circular (plane) domain.

Both of these surface domains (discs for short) are bordered by the circle k. The centre of the plane figure is C and that of the paraspherical one is A.

§18

The topic of this section is that, *in hyperbolic geometry, a plane through a point of the parasphere — provided it neither contains the axis corresponding to that point nor is perpendicular to it — intersects the parasphere in a circle.*

Actually it is only proved that if A, B, C are three points of the parasphere and the plane ABC does not contain the axis $\overrightarrow{AM}$ then ABC intersects the parasphere in a circle.

It is left an open question whether, in a plane which is not perpendicular to $\overrightarrow{AM}$ and which contains A but not M, the parasphere has two points different from A. We prove that it really has.

Let P be any point of $\overrightarrow{AM}$. Assume that the plane Γ contains A but not M and is not perpendicular to $\overrightarrow{AM}$. If Q is the foot of P in Γ and

$$\overrightarrow{QN} \| \overrightarrow{AM},$$

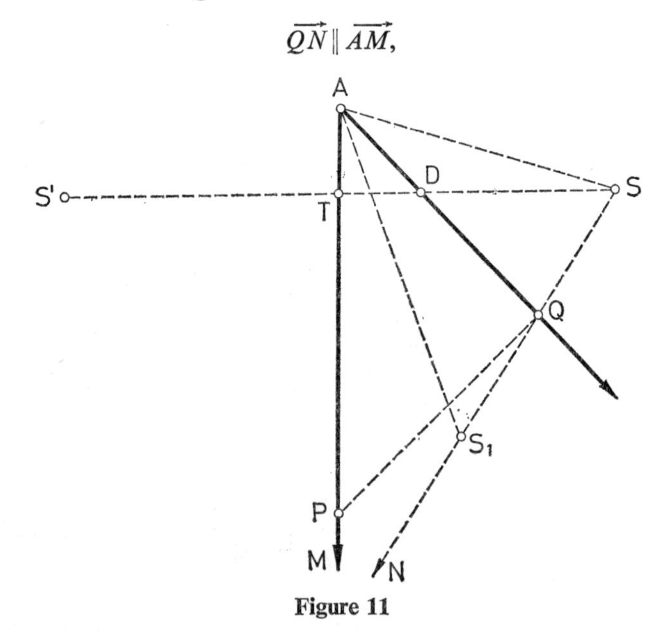

Figure 11

then by the definition of parallelism it follows that $\overrightarrow{QP}$ is in $\sphericalangle AQN$. Let S be the point isogonally corresponding to A on QN. Since

$$\sphericalangle PAQ < R < \sphericalangle AQN,$$

S does not coincide with Q. On the other hand, S cannot lie on $\overrightarrow{QN}$ either, as any point S_1 of $\overrightarrow{QN}$ satisfies $\sphericalangle AS_1N > \sphericalangle AQN > \sphericalangle MAQ > \sphericalangle MAS_1$. Thus S lies on the extension of $\overrightarrow{QN}$.

Let T be the foot of S on $\overrightarrow{AM}$. Since the isogonal secant creates acute angles with the parallel lines, T is on $\overrightarrow{AM}$. Thus Γ separates S from T and, all the more, separates S from its reflection S' in AM.

If we let S rotate about AM, it describes a circle s which lies on the parasphere and contains S'. Since Γ separates S and S', the circle s intersects Γ at two points, say B and C. These are common points of the parasphere and Γ.

On the parasphere we thus have three points A, B and C that span the plane Γ, where Γ is not incident with AM. This provides the necessary supplement to the proof given in **§18**.

Moreover, from symmetry considerations it follows immediately that the point A', diametrically opposite to A on the circle cut out by Γ, belongs to $\overrightarrow{AQ}$ and the distance of parallelism corresponding to the angle of parallelism MAQ is $\frac{1}{2}\overline{AA'}$.

§19

This section says that, *in hyperbolic geometry, the axis intersects the paracycle or the parasphere perpendicularly.*

Thus, in view of what we have seen, the parasphere and a plane can be related in four ways. Either the plane does not intersect the parasphere, or is tangent to it, or intersects it in an ordinary circle, or (the principal plane) intersects it in a paracycle.

§20

Two points of the parasphere can be connected by one and only one paracycle arc (on the parasphere).

The angle of intersection of two paracycles on a parasphere is measured by the dihedral angle of their planes. In the case of hyperbolic geometry, this relation between angles is illustrated by Fig. 10. The angle formed by the arcs $\overset{\frown}{AB}$ and $\overset{\frown}{AB'}$ on the parasphere is equal to the angle formed by the segments $\overline{CB}$ and $\overline{CB'}$ in the plane of the circle k.

The theorem of this section is valid in absolute geometry once again; also in the sequel — unless it is explicitly stated that the notions and relations are meant in System Σ or System S — all relations will be those of absolute geometry.

§21

According to this section, *if two paracycles of the parasphere are intersected by a third one and the interior angles they form with the third paracycle on one side of the latter have sum* $<2R$, *then the first two paracycles meet each other at a point on that side.*

Leaning on this theorem, the *Appendix* asserts that Euclidean geometry is valid on the parasphere provided the role of straight lines is taken over by paracycles. To see this rigorously one has to verify that all axioms of Euclidean plane geometry are really satisfied in this model.

Here we do not give all details. We mention that **§20** guarantees precisely the validity of the axiom which states that, given two points, one and only one straight line can be drawn through them. As to the axioms of congruence, their validity follows from the fact that the parasphere can be turned into itself about any of its axes, and that by suitable choice of the rotation any point of the parasphere can be taken into a given point of it; further that the parasphere is symmetric with respect to any of its principal planes.

§22

If we shift the paracycle within itself and regard the axes as if rigidly bound to the curve, then any point of the axes describes a paracycle. We say the paracycles generated in this way are mutually *parallel* or *equidistant.* Parallel paracycles have all their normals in common and the distance between two paracycles measured along any normal is the same.

All paracycles are congruent.

If we attach the paracycle to its axis rigidly and shift the axis within itself, then the paracycle replaces a parallel paracycle; in the case of hyperbolic geometry its other axes replace different axes.

Between the points of two parallel paracycles a one-to-one correspondence can be defined by associating the two points of intersection of the common normals. Two arcs of two equidistant paracycles are called *corresponding arcs* if their endpoints correspond to each other in the sense just specified.

§23

The theorems of this section can be stated also in the following form.

The quotient of two arcs of the paracycle is equal to the quotient of the corresponding arcs of any paracycle parallel to the original one.

The quotient of two corresponding arcs of two parallel paracycles depends only on the distance between the paracycles.

For the time being, we do not even know whether the function $X=f(x)$ really varies with x or is a constant that does not depend on x either. All this is clarified in the next section.

§24

This section discusses the properties of the function $X=f(x)$. In Euclidean geometry we have $f(x)=1$, that is, X does not depend on x. In hyperbolic geometry, however, — if in the denominator of the quotient of arcs defining X we consider always that arc of the paracycle which lies in the direction marked out by the paracycle axes — we have $X>1$, and if $x=y$ then also $X=Y$, while if $x<y$ then $X<Y$; the exact interdependence between x and X is expressed by the functional equation

$$X^y = Y^x.$$

By which functions is this functional equation satisfied? We get an answer to this question only in §30.

The last paragraph of this section is concerned with the congruence of strips bounded by parallel lines. It is a substantial difference between the strip of Euclidean and hyperbolic geometries that in Euclidean geometry there exist *non-congruent* strips whereas *in hyperbolic geometry any two strips are congruent (Fig. 12).*

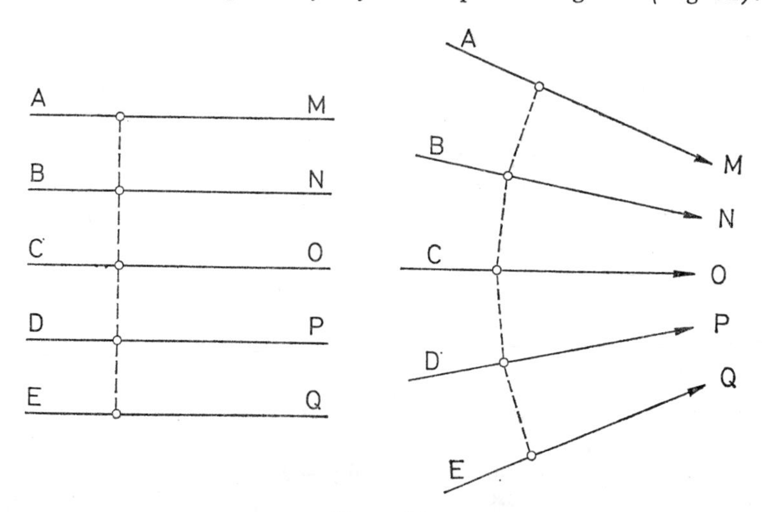

Figure 12

Thus if, in Euclidean geometry, $AM \| BN$ and the reflection of AM in BN is CO, further the reflection of BN in CO is DP, and the reflection of CO in DP is EQ, then all the strips

$$(AM, BN), (BN, CO), (CO, DP), (DP, EQ)$$

are congruent. But, for instance, the strips

$$(AM, EQ) \quad \text{and} \quad (AM, BN)$$

are non-congruent. We might say in this sense that *"the part is not congruent with the whole"*.

In hyperbolic geometry, besides the congruence of the elementary strips arising from $\overrightarrow{AM} \| \overrightarrow{BN}$ by reflection in the manner described above, also the theorem

$$(AM, EQ) \equiv (AM, BN)$$

is true. In this sense, now *"the part is congruent with the whole"*.*

To help understanding the problems considered above, we present some additional theorems.

A. *In hyperbolic geometry the angle sum of the triangle is less than* $2R$.

For let $\sphericalangle ACB = R$ and let $\overrightarrow{AM}, \overrightarrow{BN}$ and $\overrightarrow{CP}$ be perpendicular to the plane ABC. Then the sum of the dihedral angles of the prism determined by the edges

$$\overrightarrow{AM}, \overrightarrow{BN}, \overrightarrow{CP}$$

is equal to the sum

$$\sphericalangle A + \sphericalangle B + \sphericalangle C,$$

that is, to the angle sum of the triangle ABC *(Fig. 13)*.

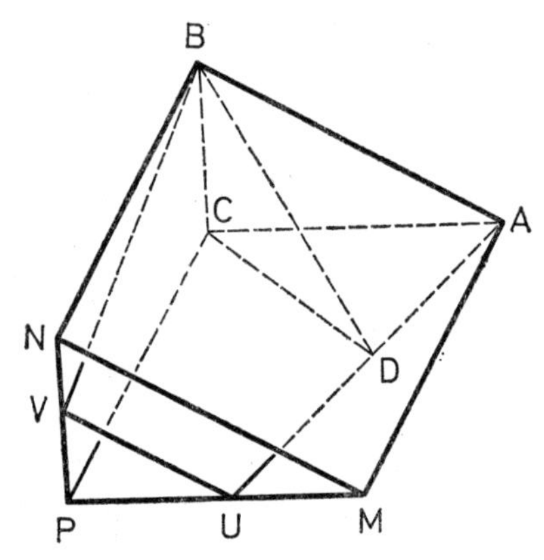

Figure 13

* This theorem sounded strange in BOLYAI's time. People were inclined to carry over without criticism such theorems as "*the part cannot be congruent with the whole*", valid for a finite domain, to infinite domains. Just the confused and uncritical attitude towards questions of a similar nature prevented contemporaries from admitting and understanding the correctness of more exact views like those of BOLYAI and LOBACHEVSKY.

We now increase this sum of dihedral angles by turning the half-plane $|AB|N$ about AB. Let

$$\overrightarrow{AU} \parallel \overrightarrow{CP},$$

and let $|AB|U$ intersect $|CB|N$ in $\overrightarrow{BV}$. Then

$$\overrightarrow{BV} \parallel \overrightarrow{CP}$$

and therefore

$$\overrightarrow{AU} \parallel \overrightarrow{BV}.$$

Now the sum of the dihedral angles of the prism formed by the parallel half-lines

$$\overrightarrow{AU},\ \overrightarrow{BV},\ \overrightarrow{CP}$$

is — according to **§21** — just $2R$. The former and the latter prisms have their dihedral right angle in common. Moreover, the dihedral angle at $\overrightarrow{AM}$ of the former prism has been replaced by the dihedral angle at $\overrightarrow{AU}$ of the latter and, consequently, this dihedral angle has increased. Really, if $CD \perp DA$ then, in view of the relation $BC \perp \perp ACP$, also $BD \perp DA$. Therefore $\sphericalangle BDC$ exhibits just the dihedral angle in question. On the other hand, the right triangles ABC and DCB have the common leg BC, and for the legs not in common we see from the right triangle ADC that

$$CD < CA.$$

Thus

$$\sphericalangle CDB > \sphericalangle CAB.$$

By a similar reasoning, the dihedral angle between $|AB|V$ and $|CB|V$ is bigger than $\sphericalangle B$.

As a result, the sum of the dihedral angles for the second prism is greater than the sum of the dihedral angles for the first prism, that is,

$$\sphericalangle A + \sphericalangle B + \sphericalangle C < 2R.$$

B. *In hyperbolic geometry to a greater distance there corresponds a smaller angle of parallelism and to a smaller angle of parallelism there corresponds a greater distance.*

Really, if

$$AB < AC$$

while

$$\overrightarrow{BN} \parallel \overrightarrow{AM} \quad \text{and} \quad \sphericalangle ABN = \sphericalangle ACO,$$

then — the sum of any two angles of a triangle being $< 2R$ — $\overrightarrow{CO}$ cannot intersect $\overrightarrow{BN}$. By **§13**, however, neither can $\overrightarrow{CO}$ be parallel to $\overrightarrow{BN}$. Thus if

$$\overrightarrow{CP} \parallel \overrightarrow{BN}$$

then $\overrightarrow{CP}$ lies within $\sphericalangle BCO$. Therefore

$$\sphericalangle BCP < \sphericalangle ACO.$$

10*

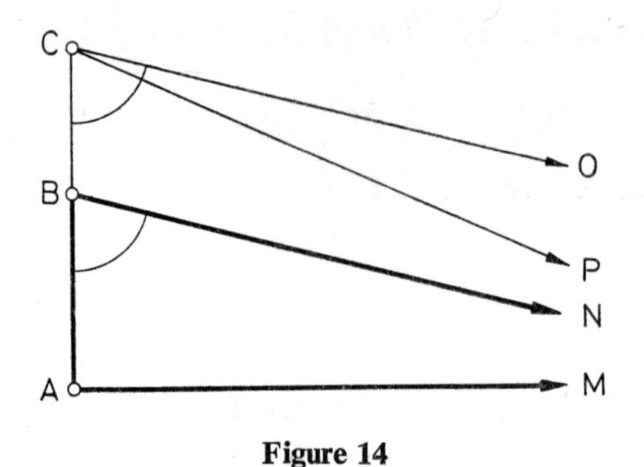

Figure 14

Further, according to §7, $\overrightarrow{CP}$ is parallel also to $\overrightarrow{AM}$. For the angles of parallelism corresponding to the distances AB and AC it now follows that

$$\sphericalangle ABN > \sphericalangle ACP,$$

as we have claimed.

C. *In hyperbolic geometry the distance of parallelism decreases in the direction of parallelism.**

For let

$$\overrightarrow{JY} \parallel \overrightarrow{AX}$$

and

$$JA \perp AX, \quad KC \perp AX.$$

Let also

$$AB = BC.$$

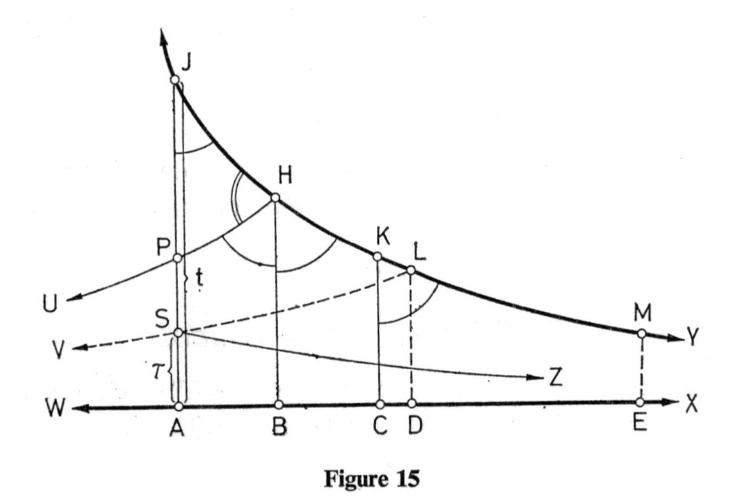

Figure 15

* We have made perceptible the asymptotic approach of the parallel lines of hyperbolic geometry to each other by a distorted diagram (*Fig. 15*).

The perpendicular to AX at B must leave the quadrilateral $AJKC$ at some point H which lies between the endpoints of the line segment JK.

If P satisfies the condition $AP=CK$, then

$$ABHP \equiv CBHK$$

and, consequently,

$$\overrightarrow{HP} \parallel \overrightarrow{BW}.$$

But the angle of parallelism in hyperbolic geometry is an acute angle; therefore $\sphericalangle JHP > 0$, and hence it already follows that

$$AJ > AP.$$

Of course, though we know that the distance of parallelism decreases in the direction of X and that $\overrightarrow{JY}$ does not step to the side of AX opposite to J, we still do not know what the limit of that decreasing distance is.

D. *The distance of parallelism* (between two parallel lines) *becomes smaller than any distance — however small — given in advance.*

Indeed, if the distance τ given in advance is smaller than $AJ=t$, lay it on $\overrightarrow{AJ}$ starting from A, and from the endpoint S thus obtained let

$$\overrightarrow{SZ} \parallel \overrightarrow{AX}.$$

Then

$$\overrightarrow{SZ} \parallel \overrightarrow{JY}.$$

Let $\overrightarrow{SV}$ be the reflection of $\overrightarrow{SZ}$ in $\overrightarrow{AJ}$. Its complementary half-line enters JSZ and, consequently, intersects $\overrightarrow{JY}$ at a point L.

Let D be the foot of L on $\overrightarrow{AX}$. Since both $\overrightarrow{LY}$ and $\overrightarrow{LV}$ are parallel to the line AE, it follows that $\overrightarrow{LD}$ is the angle bisector of SLM. The symmetry of the figure implies that if $AD=DE$ then for the perpendicular EM drawn to AE at E

$$EM = AS = \tau.$$

A more intuitive description of the theorem just proved is the following. The strips (AX, JY) and (AX, SZ) are congruent: the latter can be shifted along $\overrightarrow{AX}$ so as to coincide with the former by shifting the segment AS onto the segment EM. This sheds new light on our remark concerning the left-hand portion of Fig. 12.

§25

In the *Appendix* it is the *absolute sine law** which, besides **§10**, shows most tipically how absolute geometry unites the theorems of Euclidean and hyperbolic geometries. §§ **10** and **25** play the most important role in the whole work.

* BONOLA, in his book *Die nichteuklidische Geometrie* (1908) writes: "Ein absoluter Satz von Bolyai, der von wunderbarer Einfachheit und Eleganz ist...".

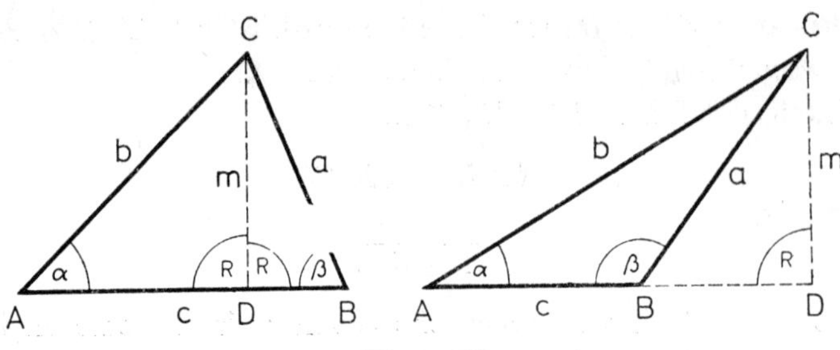

Figure 16

If we apply the theorem proved in **§25**, say, to the right triangles *ADC* and *BDC* determined by the altitude to side c, we obtain

$$\circ m : \circ a = \sin\alpha : 1, \quad \circ b : \circ m = 1 : \sin\beta.$$

Therefore, denoting the angles opposite to sides a, b and c of the general triangle by α, β and γ, respectively,

$$\circ a : \circ b : \circ c = \sin\alpha : \sin\beta : \sin\gamma.$$

Hence, making use of the expression $2\pi r$ for the circumference of the circle of radius r, we immediately arrive at the sine law of Euclidean geometry:

$$a : b : c = \sin\alpha : \sin\beta : \sin\gamma.$$

To deduce the sine law of hyperbolic geometry we need the expression for $\circ r$ in terms of r, which by **§30** is the following:

$$\circ r = 2\pi k \operatorname{sh}\frac{r}{k}.$$

Consequently, the sine law assumes the form

$$\operatorname{sh}\frac{a}{k} : \operatorname{sh}\frac{b}{k} : \operatorname{sh}\frac{c}{k} = \sin\alpha : \sin\beta : \sin\gamma.$$

This theorem exhibits strong formal relationship with the sine law of spherical trigonometry. As it is well known, if α, β, γ are the angles of the spherical triangle, while a, b, c and r are the lengths of its sides and the radius of the sphere, then

$$\sin\frac{a}{r} : \sin\frac{b}{r} : \sin\frac{c}{r} = \sin\alpha : \sin\beta : \sin\gamma.$$

§26

According to this section, the relations of spherical trigonometry can be deduced from the residual system of axioms or, in other words, they are absolute theorems. However, treated is only the sine law for the spherical right triangle in the case where the angle measures of the legs are smaller than the right angle. We show that, in fact, the theorems for the general spherical triangle can be derived from this result.

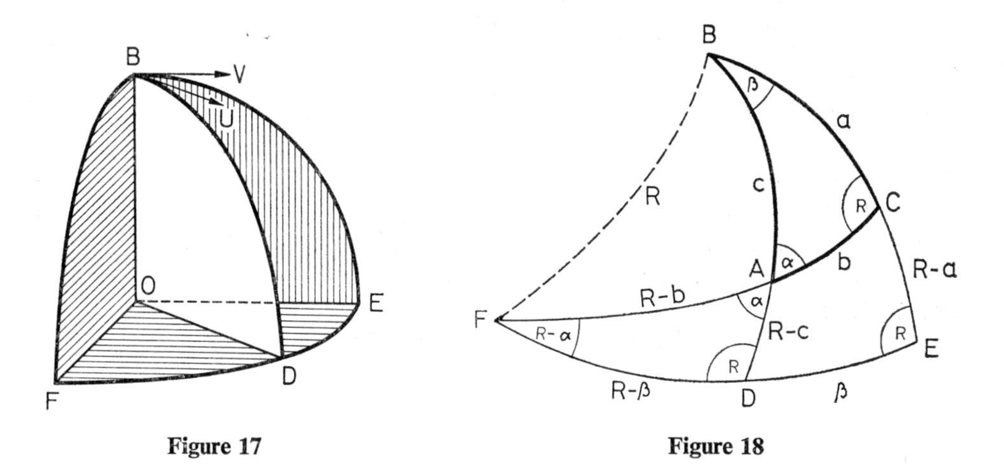

Figure 17 Figure 18

We first prove the theorems for the spherical right triangle.

If in a spherical triangle $\gamma = R$, *then*

1°. $$\sin a = \sin \alpha \sin c, \quad \sin b = \sin \beta \sin c,$$

2°. $$\cos \alpha = \cos a \sin \beta, \quad \cos \beta = \cos b \sin \alpha,$$

3°. $$\cos c = \cos a \cos b.$$

We start with the case $a < R$, $b < R$. Then 1° is proved in the *Appendix*. To establish 2° and 3° we consider a spherical octant BEF all sides and angles of which are equal to R (Fig. 17).

Fitting the spherical triangle ABC to vertex B of the octant we obtain the dissection shown by Fig. 18. Applying relations 1°, already proved, to the spherical triangle ADF

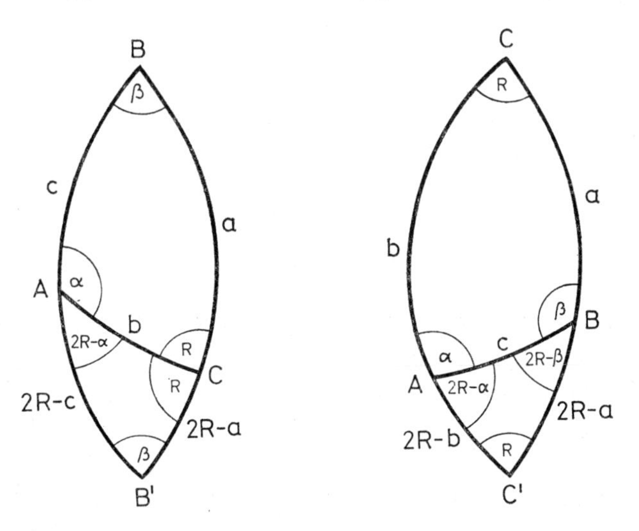

Figure 19

we arrive at 3° and the second equation of 2°. By reasons of symmetry also the first equation of 2° is valid.

The validity of our assertion for general spherical right triangles follows from the fact that if one or both of the legs are greater than R then, completing the spherical triangle to a spherical lune as shown by Fig. 19, the legs of the complementary spherical triangle are smaller than R. Our equations applied to these complementary spherical triangles yield just the relations stated for the original spherical triangles.

We now prove *the sine law and the cosine laws* for general spherical triangles (Figures 20-21):

$$\sin a : \sin b : \sin c = \sin \alpha : \sin \beta : \sin \gamma,$$

$$\cos a = \cos b \cos c + \sin b \sin c \cos \alpha,$$

$$\cos \alpha = -\cos \beta \cos \gamma + \sin \beta \sin \gamma \cos a.$$

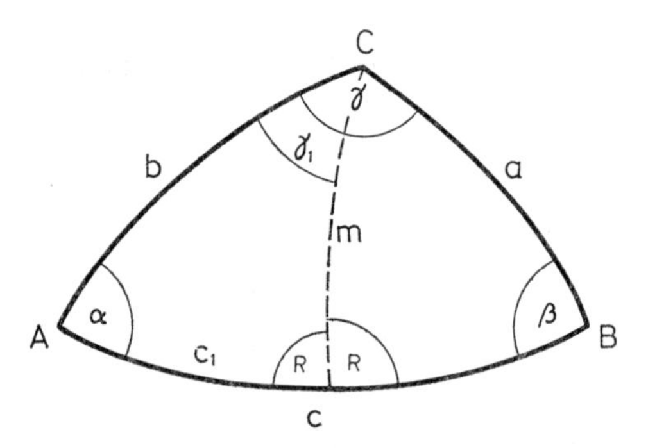

Figure 20

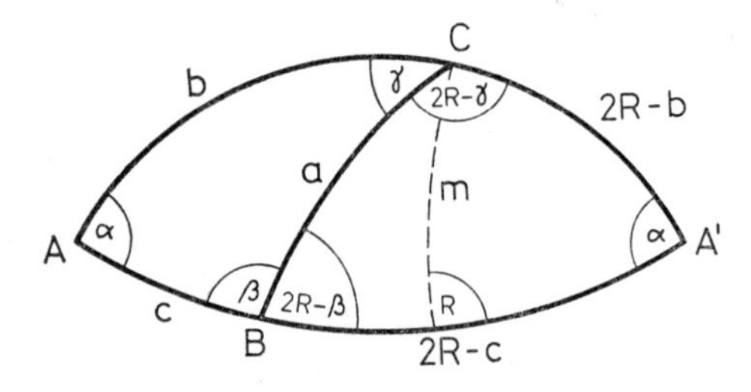

Figure 21

We dissect the spherical triangle in two others by the altitude m that starts from the vertex of γ. If this altitude does not divide the original spherical triangle into two parts, then we first carry out the proof concerning that spherical triangle for which a dissection actually occurs and which completes our spherical triangle to a spherical lune.

For the two spherical right triangles, according to what has already been proved,

$$\sin m = \sin \alpha \sin b = \sin \beta \sin a$$

and the sine law follows. Again by the relations proved above

$$\sin c_1 = \sin \gamma_1 \sin b, \quad \cos \alpha = \cos m \sin \gamma_1,$$
$$\cos b = \cos m \cos c_1, \quad \cos a = \cos m \cos (c-c_1).$$

Consequently,

$$\cos a = \cos m \cos c_1 \cos c + \cos m \sin c \sin c_1 =$$
$$= \cos b \cos c + \frac{\cos \alpha}{\sin \gamma_1} \sin c \sin \gamma_1 \sin b =$$
$$= \cos b \cos c + \sin b \sin c \cos \alpha.$$

We have thus proved the cosine law for sides; the cosine law for angles hence follows with the help of the polar spherical triangles in the well-known manner.

§27

This section deals with deducing the formula

$$\sin u : \sin v = 1 : \sin z$$

by means of the distance line and distance surface*. So the point in question is that if the quadrilateral $ABCD$ satisfies the conditions

$$AC = BD, \quad \sphericalangle CAB = \sphericalangle DBA = R$$

* In Euclidean geometry, the distance lines to a given straight line are the lines parallel to the given one. This property of straight lines (namely that distance lines are straight lines) is characteristic for Euclidean geometry. The concept of distance line came into prominence when the vain attempts at deducing the Euclidean axiom of parallelism from the residual system of axioms started. Farkas Bolyai, for instance, has tried to deduce from the residual system of axioms that the distance line is a straight line; from the latter circumstance the angle relation expressed by the Euclidean axiom of parallelism can be deduced as a theorem.

Apparently, JÁNOS BOLYAI has also probed this way, but he discovered that the distance line being a straight line is equivalent to the Euclidean axiom of parallelism. Gauss has also meditated on this question; as a matter of fact, he raised and solved the following problem concerning curved surfaces of Euclidean space.

Consider a curved surface which has the following property: there is a simply connected region in the surface — possibly coinciding with the whole surface — any two points of which can be connected by one and only one *g-arc* (geodesic arc). By a *g*-arc we mean a curve in the surface for which the shortest path in the surface between any two of its points is just the part of the *g*-line between these points. In this surface, starting at right angles from all points of a *g*-line, let us draw *g*-arcs of one and the same length. Their endpoints may form a line in the surface and then we call it a distance line of the surface. One may ask whether the distance line so defined is in general a *g*-line. It is not. (An easy example is the sphere. A *g*-line of the sphere is a great circle. On the other hand, the distance line of a great circle is in general a small circle of the sphere and is therefore not a *g*-line.) Then the following and the diagonal AD forms angles u and v with the sides AC and BD respectively, further if z is the angle of parallelism corresponding to the distance BD, then the interdependence of u, v and z is expressed by the relation above. By the way there also appears a theorem according to which the ratio of the segment $\overline{AB}$ of the base line and the corresponding arc of the distance line is a function of the parameters u and v, namely

$$\overset{\frown}{CD}:\overline{AB} = \sin u : \sin v.$$

We give a short discussion of fundamental facts concerning the distance line and distance surface.

A. The locus of all points at distance d from line a on one side of the latter is called distance line (T-line) at distance d for the base line a.

The locus of all points at distance d from the plane α on one side of the latter is called distance surface (τ-surface) at distance d for the base plane α.

The lines which are perpendicular to the base line or base plane are called axes of the distance line or distance surface, respectively.

B. By the simplest congruence considerations it follows that letting the distance line rotate about any of its axes we obtain the distance surface to the same distance. During rotation, the base line of the distance line sweeps the base plane of the distance surface in question. The distance surface is a surface of revolution with respect to any of its axes. The planes which are perpendicular to the base plane intersect the distance surface in congruent distance lines. Any plane perpendicular to the base plane is a plane of symmetry for the distance surface. The distance line is symmetric with respect to any of its axes and, as a simple consequence, can be shifted within itself.

C. *From the assumption that a certain distance line contains three points of a straight line the Euclidean axiom of parallelism can be derived*; in this case any three points of any distance line are collinear. Thus, assuming that on a certain distance line three non-collinear points can be found, the theorem according to which the angle of parallelism is $<R$ can be deduced; then, consequently, any three points of any distance line are non-collinear. (Inducing theorems.)

We base the proof on the theorem which says that from the existence of a rectangle the system Σ can be deduced (since the diagonal divides the rectangle into triangles with angle sum $=2R$ each).

---

question arises: does there exist a surface all distance lines of which are g-lines? As proved by Gauss, only the surfaces of curvature 0 enjoy this property. In other words, only the developable surfaces. Since the plane has curvature 0, its distance lines are g-lines. Thus we have come back to Farkas Bolyai's problem. The concept of curvature and those theorems on curvatures which are deeply exploited in Gauss' demonstration are not independent of the Euclidean axiom of parallelism.

The work of János Bolyai shows clearly that the attempts of his father Farkas to find a proof could not have succeeded. The line of constant distance is a straight line in Euclidean geometry and a uniform curve in hyperbolic geometry.

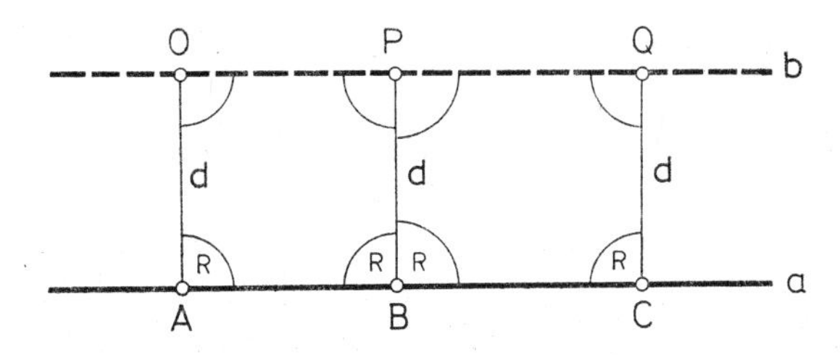

Figure 22

So let the endpoints O, P, Q of the perpendiculars of length d at points A, B, C of the line a belong to line b. From the symmetry of the quadrilaterals $ABPO$, $ACQO$ and $BCQP$ (Fig. 22)

$$\sphericalangle BPO = \sphericalangle AOP = \sphericalangle CQP = \sphericalangle BPQ.$$

Since, however, $\sphericalangle BPQ$ and $\sphericalangle BPO$ are supplementary angles, each of them is a right angle and the quadrilaterals mentioned above are rectangles.

The distance line and distance surface exhibit further interesting properties just in System **S**, since there **T** and τ are a uniform curve and a uniform curved surface, respectively.* The next theorems of this section hold true in System **S**.

D. *If a and b are non-intersecting non-parallel lines, then there is one and only one line s that intersects each of them at a right angle.*

We prove the theorem through continuity.** From **§1** it follows that there exists a right triangle in which one leg and the acute angle opposite to it are given arbitrarily. In other words, given an acute angle, on one of its arms there is a point whose perpendicular distance from the other arm is prescribed arbitrarily.

If we intersect the arms of a right angle by lines which form a given angle with one arm, then these secant lines do not intersect each other; indeed, the angle sum of a triangle cannot be greater than 180°. This and the fact stated precedingly imply that if a point moves away from the vertex of an acute angle along one of the arms then the perpendicular distance from this point to the other arm increases monotonically and surpasses any bound.

Consider the non-intersecting non-parallel lines a and b. Any point P of line a is the origin of two half-lines p and p' parallel to b. According to what we have established so far, for each of these half-lines it is true that the perpendicular distance to a from a point moving away on them increases monotonically and indefinitely. Therefore, considering the perpendicular distance $d(B)$ from the variable point B of b to

* In the course of reading the *Appendix* we become more and more convinced that in his concise work the author presents only the most indispensable out of the abundant results of his discovery. No doubt, JÁNOS BOLYAI knew all the theorems which we next describe.

** HILBERT proved this theorem independently of the axioms of continuity (that is, relying only on the planar axioms of incidence, order, and congruence).

line a we find that $d(B)$ tends to infinity as B moves off to infinity in some direction along b.

Since $d(B)$ is a continuous function of the position of B, by the facts mentioned above this function attains its minimum at a finite point: for some point B_0 the value $d(B_0)$ is minimal. The segment A_0B_0 of length $d(B_0)$ arising in this way is perpendicular also to b; otherwise denoting by B_1 the projection of A_0 to b we would have $A_0B_1 < d(B_0)$ and all the more $d(B_1) < d(B_0)$, which is impossible.

Besides the common perpendicular A_0B_0 so obtained, the lines a and b cannot have any other common perpendicular. In fact, two perpendiculars would determine a rectangle while there is no rectangle in System **S**.

Since, for the axes $\overrightarrow{MA}$ and $\overrightarrow{NB}$ of the distance line, symmetry yields $\overrightarrow{MA} \simeq \overrightarrow{NB}$, isogonal correspondence suggests a unified definition of circle, paracycle, and hypercycle.

By a *flat pencil of lines* we mean the collection of all lines

a) through a point,

b) parallel to a line,

c) perpendicular to a line.

Take a line of a flat pencil and select a point on it. Consider the isogonal companions of this point on the other lines of the pencil. The collection of all these points forms, corresponding to the three cases listed above,

a) the circle,

b) the paracycle,

c) the hypercycle.

This definition of the hypercycle covers also the straight line (hypercycle that coincides with its base line).

In space, similarly, a unified definition of sphere, parasphere and hypersphere (plane) can be given.

E. *Three points can be connected by either a straight line or a uniform curve.**

Let A, B and C be three non-collinear points and let the perpendicular bisectors of AB, BC and CA be w, u and v, respectively. Three cases may occur.

I. u and v meet at some point O.

II. u and v form a pair of parallel lines.

III. u and v form a pair of non-intersecting non-parallel lines.

It should be noted that if A, B and C are different points then u and v cannot coincide since from C only one perpendicular can be dropped to a line and C has only one reflection in a line.

* Cf. Part I, Section 2. The axiom of FARKAS BOLYAI, which may be chosen instead of the Euclidean axiom of parallelism, says that (in System Σ) three points are on either a circle or a line. Also this basic theorem indicates the wealth of hyperbolic geometry compared with Euclidean geometry.

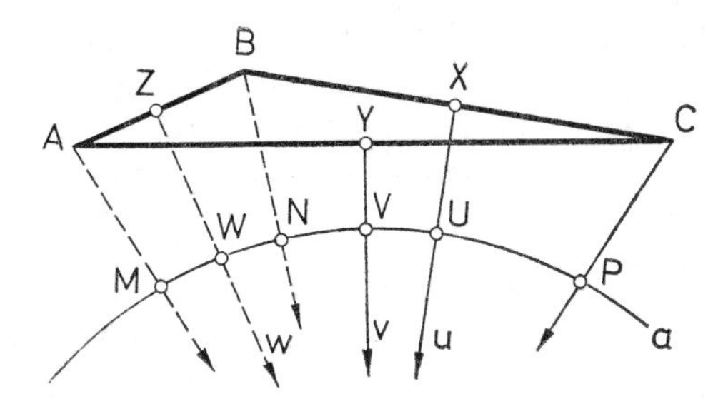

Figure 23

Case I. For the point O on the lines of symmetry of both sides it is obvious that

$$OC = OA, \quad OC = OB,$$

and therefore

$$OA = OB.$$

Consequently, the circle of centre O and radius OC passes through all vertices of the triangle ABC.

Case II. Through the point C, draw $\overrightarrow{CP}$ to be parallel to u in the direction in which u and v are parallel. Let the reflections of $\overrightarrow{CP}$ in u and v be $\overrightarrow{BN}$ and $\overrightarrow{AM}$, respectively. Then (Fig. 23)

$$\overrightarrow{BN} \| \simeq \overrightarrow{CP}, \quad \overrightarrow{AM} \| \simeq \overrightarrow{CP},$$

and therefore

$$\overrightarrow{BN} \| \simeq \overrightarrow{AM}.$$

Hence A, B and C all lie on the paracycle that belongs to point C of the axis $\overrightarrow{CP}$.

Case III. Consider the common perpendicular a to lines u and v. Drop a perpendicular from C to a. Denote its foot by P. It is obvious that the length of BN and AM, the reflections of CP in u and v is equal to CP. It is also clear that BN and AM meet a perpendicularly at N and M.

Thus the vertices of the triangle ABC are on the hipercycle that lies at distance PC from the base line a.

In conclusion we mention a few facts relating to the hypersphere without proof. They can be easily proved on the basis of what has been said so far.

Let α be the base plane, τ the hypersphere at distance d from it, M a point of this hypersphere, and A the perpendicular projection of M to α. Let the line s pass through M and be perpendicular to MA. Let σ be a plane incident with s, and consider the intersection of σ and τ.

Let us turn σ from the initial position when it passes through A (and therefore is also incident with MA) to the final position perpendicular to MA. From the initial position to the position parallel to α the intersection is a hypercycle the base line of which is the intersection of σ and α. The distance that defines the intersection increases from d to infinity. As σ takes the position parallel to α, the intersection becomes a paracycle. If we turn the plane σ further, the intersection will be a circle of decreasing radius, and in the position perpendicular to MA this circle contracts to a point. In this final position, σ is the tangent plane at point M of the surface τ.

Thus the hypersphere makes it possible to regard uniform curves, apart from hypercycles of parameter less than d, as elements of a continuous system.

The axes of the hypersphere τ to the base plane α and distance d induce a one-to-one correspondence between the points of τ and α (associated with each point of τ is its perpendicular projection to α).

This correspondence associates the hypercycles of minimal parameter (that is, of parameter d) in τ with the straight lines in α. The shortest path between points M and N of τ within the surface is the hypercycle arc the plane of which is perpendicular to α. The intersections of the surface with the planes perpendicular to α will be called — by analogy with the principal (great) circles of the sphere — the *principal lines* (or principal hypercycles) of τ.

The angle formed by two principal lines is equal to the angle formed by the respective base lines (projections of the principal lines to plane α). Thus the correspondence in question *preserves angles, bends* the straight lines of the base plane and, according to the formula which appears at the beginning of this section, *extends* them by the factor

$$\lambda = \sin u : \sin v.$$

if on the hypersphere the role of straight lines is played by principal lines, then hyperbolic geometry is valid also on the hypersphere.

§28

A theorem similar to the theorem of this section holds for the quotient of corresponding arcs of concentric circles. In fact (Fig. 24),

$$\circ x : \circ y = \overset{\frown}{AB} : \overset{\frown}{CD}$$

and, on the other hand, for the triangle BCO in view of §25

$$\circ x : \circ y = \sin u : \sin v.$$

Consequently

$$\overset{\frown}{AB} : \overset{\frown}{CD} = \sin u : \sin v.$$

If we replace $\mathbf{K}'$ by a straight line and $\mathbf{K}$ by a hypercycle at constant distance from this line, then according to §27

$$\overset{\frown}{AB} : \overline{CD} = \sin u : \sin v.$$

If we replace both of $\mathbf{K}$ and $\mathbf{K}'$ by hypercycles which enclose a curved strip of constant width, then the theorem can be extended also to the quotient of their arcs. For,

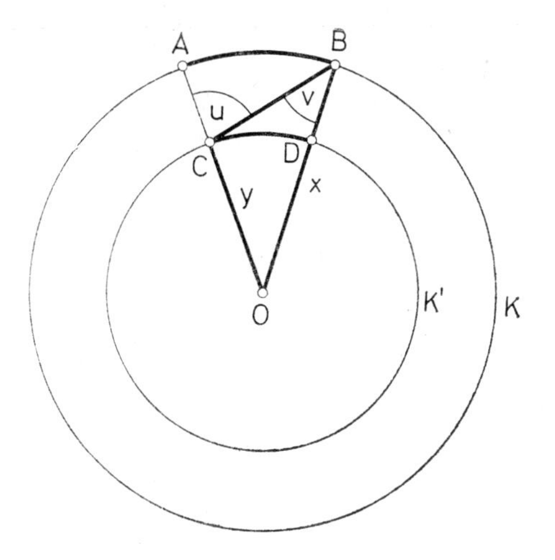

Figure 24

applying the theorem of **§27** to the hypercycles $\widehat{AB}$ and $\widehat{CD}$ at distances x and y from the line PQ (Fig. 25) it follows that

$$\widehat{AB}:\overline{PQ} = \sin u':\sin v'$$

and

$$\widehat{CD}:\overline{PQ} = \sin u'':\sin v.$$

Hence

$$\widehat{CD}:\widehat{AB} = \frac{\sin u'' \cdot \sin v'}{\sin u' \cdot \sin v''}.$$

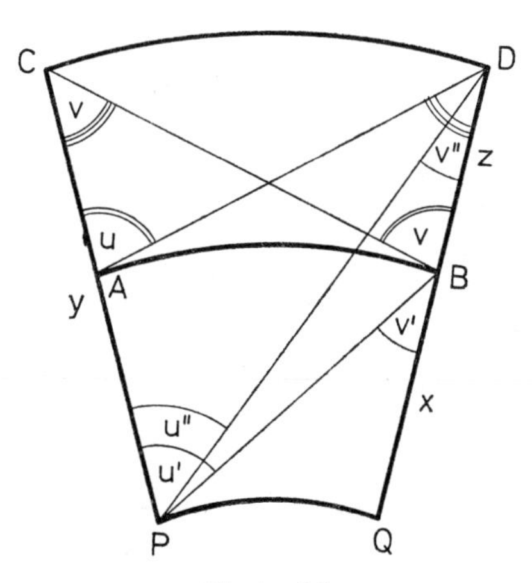

Figure 25

According to §25 applied to the triangles ADP, BCP and BDP

$$\circ PD:\circ AD = \sin u:\sin u'',$$

$$\circ CB:\circ PB = \sin u':\sin v,$$

$$\circ PB:\circ PD = \sin v'':\sin v'.$$

Forming the product and taking into account the relation $AD=BC$ we obtain

$$\frac{\sin u}{\sin v} = \frac{\sin u''\cdot\sin v'}{\sin u'\cdot\sin v''},$$

that is,

$$\widehat{CD}:\widehat{AB} = \sin u:\sin v.$$

Summing up: *the quotient of the corresponding arcs of two equidistant uniform lines is, in any case, given by*

$$\lambda = \sin u:\sin v.$$

§29

This section deals with the planimetric deduction of the relation between the distance of parallelism and the corresponding angle of parallelism in System S. A stereometric derivation of the relations proved here is given, for instance, in the paper of J. KÜRSCHÁK: On the angle of parallelism [Math. Phys. Lapok **12** (1903), 50–52 (in Hungarian)].

§30

In this section the perimeter of the circle is determined without integration. It is proved that

$$\circ y = 2\pi k\,\mathrm{sh}\,\frac{y}{k}.$$

It is also proved in the section that for the angle of parallelism* $\Pi(y)=R-z$ corresponding to distance y we have

$$\mathrm{ctg}\,\Pi(y) = \mathrm{sh}\,\frac{y}{K}.$$

On the other hand, the ratio of corresponding arcs of two paracycles at mutual distance y equals

$$Y = e^{\frac{y}{k}}.$$

* The notation $\Pi(x)$ for the angle of parallelism to distance x goes back to LOBACHEVSKY.

By **§27** the ratio of corresponding pieces of the distance line to distance y and the base line is

$$\mathrm{E}(y) = \frac{1}{\sin \Pi(y)} = \operatorname{ch} \frac{y}{k}.$$

This section contains a part which refers merely to intuition, namely the assertion

$$\frac{r}{y} \to 1 \quad \text{as} \quad y \to 0,$$

that is, the theorem according to which *the quotient of the paracycle arc and its chord tends to* 1 *when the arc tends to* 0. In the *Appendix*, the existence of arc length appears on an intuitive basis.

We fill this gap in the following way. According to **§16** the paracycle is a convex line since a straight line intersects it in no more than two points. Consequently, the lengths of the broken lines inscribed in a paracycle arc tend to a limit as the subdivision of the arc is refined indefinitely. This limit provides the arc length of the paracycle. We thus have established a connection between the arc length defined by means of polygons and the arc length defined for **L**-lines on the parasphere according to **§21**.

Since the paracycle arc is a uniform line and the length of an inscribed polygon made up of distances of length $2y$ tends to the arc length of the paracycle as y decreases, for the paracycle arc having length $2r$ and corresponding to a chord of length $2y$ (making use of the monotonicity not explained here in detail)

$$\lim_{y \to 0} \frac{r}{y} = 1.$$

From this statement it follows that the definitions of arc length by polygons or by **L**-polygons (for instance, in determining the perimeter of the circle) lead to the same result.

In **§30** it is also proved that for the chord of length $2y$ belonging to the paracycle arc of length $2r$

$$\frac{r}{k} = \operatorname{sh} \frac{y}{k},$$

which in turn yields the limiting relation we have used in deducing it and proved above. From this formula with the help of the expression of $\Pi(y)$ it can be seen that *k is the half-length of the paracycle arc to the half-chord of which the angle of parallelism is* 45°.

§31

In this section the following formulas concerning right triangles are deduced:

$$\operatorname{sh}\frac{a}{k} = \operatorname{sh}\frac{c}{k}\sin\alpha, \tag{1}$$

$$\cos\alpha = \operatorname{ch}\frac{a}{k}\sin\beta, \tag{2}$$

$$\operatorname{ch}\frac{c}{k} = \operatorname{ch}\frac{a}{k}\operatorname{ch} bk. \tag{3}$$

Further the relation $\operatorname{sh}^2\frac{c}{k}=\operatorname{ch}^2\frac{a}{k}\operatorname{sh}^2\frac{b}{k}+\operatorname{sh}^2\frac{a}{k}$, which could be obtained directly from (3), is established.

Similarly to (1) and (2) we have

$$\operatorname{sh}\frac{b}{k} = \operatorname{sh}\frac{c}{k}\sin\beta, \tag{1'}$$

$$\cos\beta = \operatorname{ch}\frac{b}{k}\sin\alpha. \tag{2'}$$

From the relations listed so far we easily obtain the formula

$$\operatorname{ctg}\alpha\operatorname{ctg}\beta = \operatorname{ch}\frac{c}{k}$$

also proved in the *Appendix*.

We call attention to the complete analogy between this theorem and those appearing in our remarks to **§26** and relating to the spherical right triangle. The analogy comprises also the sine law of **§25**. By exactly following the argument of the remarks to **§26** we obtain the cosine law of hyperbolic geometry:

$$\operatorname{ch}\frac{a}{k} = \operatorname{ch}\frac{b}{k}\operatorname{ch}\frac{c}{k}+\operatorname{sh}\frac{b}{k}\operatorname{sh}\frac{c}{k}\cos\alpha.$$

The only thing to be noted is that in the case of an obtuse triangle now we cannot work with the triangle completed to a lune; instead, we apply the original proof to the right triangle which contains and completes the given one. This causes the only change that $c-c_1$ is replaced by c_1-c, which does not affect the result since ch is an even function.

Also comprised by the above-mentioned analogy is the cosine law for angles in hyperbolic trigonometry:

$$\cos\alpha = -\cos\beta\cos\gamma+\sin\beta\sin\gamma\operatorname{ch}\frac{a}{k}.$$

Its proof, however, cannot be based on polar triangles. Instead, we proceed as follows.

With the notation of Fig. 26 and using the facts stated above

$$\operatorname{ch}\frac{a}{k} = \operatorname{ch}\frac{c_2}{k}\operatorname{ch}\frac{m}{k}, \quad \cos\gamma_2 = \operatorname{ch}\frac{c_2}{k}\sin\beta,$$

$$\cos\beta = \operatorname{ch}\frac{m}{k}\sin\gamma_2, \quad \cos\alpha = \operatorname{ch}\frac{m}{k}\sin(\gamma-\gamma_2).$$

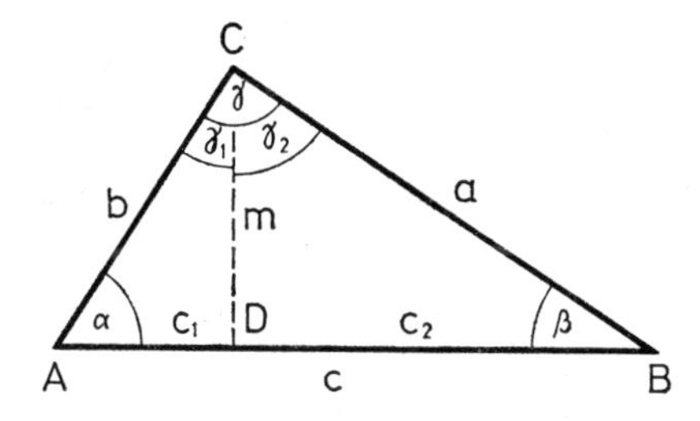

Figure 26

Hence

$$\cos\alpha = -\operatorname{ch}\frac{m}{k}\cos\gamma\sin\gamma_2 + \operatorname{ch}\frac{m}{k}\sin\gamma\cos\gamma_2 =$$

$$= -\cos\beta\cos\gamma + \frac{\operatorname{ch}\frac{a}{k}}{\operatorname{ch}\frac{c_2}{k}}\sin\gamma\cdot\operatorname{ch}\frac{c_2}{k}\sin\beta,$$

which gives the assertion. Again, it causes no difficulty if α is obtuse since it is sufficient then to replace $\gamma-\gamma_2$ by $\gamma_2-\gamma$ and, the cosine function being even, this does not alter the result.

For $k\to\infty$ all relations treated above turn into theorems of Euclidean trigonometry. We point out that Theorem III turns in this way into the Pythagorean theorem. The Euclidean theorem is immediately obtained by taking limits in the relation

$$\circ a^2 + \circ d^2 = \circ c^2,$$

BOLYAI's Pythagorean theorem of absolute geometry.

§32

In this section, BOLYAI deals with determining the measure of geometric figures by analytic tools. He uses a rectangular system of coordinates in the plane where the coordinates of a point are: the perpendicular distance y of the point from the x-axis, and the distance x of the foot of this perpendicular from the origin. This must be emphasized because in the hyperbolic plane e.g. the perpendicular distance of the point from the y-axis is not equal to x.

The contents of the section are the following.

I. The direction number of the curve $y=f(x)$, that is, the tangent of the angle formed by the tangent of the curve and the perpendicular to the ordinate of the point considered is given by the formula

$$\operatorname{tg}\alpha = \lim_{\Delta y \to 0} \frac{\Delta y}{\Delta t},$$

where Δt denotes the length of that arc of the distance line passing through the point and having the x-axis for base line which corresponds to the portion Δx of the x-axis.

II. For the arc length Δz of an arc of the curve $y=f(x)$ we have

$$\frac{(\Delta z)^2}{(\Delta y)^2+(\Delta t)^2} \to 1.$$

The equation of the paracycle that touches the y-axis at the origin is

$$\operatorname{ch}\frac{y}{k} = e^{\frac{x}{k}},$$

and its arc length from the origin to the point equals

$$z = \sqrt{e^{\frac{2y}{k}}-1}.$$

III. The area bounded by the curve $y=f(x)$ in the interval (a, b) is given by

$$u = k\int_a^b \operatorname{sh}\frac{y}{k}\,dx.$$

The ratio of the area of a region in the distance surface (hypersphere) to distance q and the area of the corresponding region in the base plane is

$$\operatorname{ch}^2\frac{q}{k}.$$

The volume of the space region enclosed by a region of area p in the plane, the perpendiculars erected on its boundary, and the distance surface to distance q is equal to

$$p\left[\frac{k}{4}\operatorname{sh}\frac{qk}{2}+\frac{q}{2}\right].$$

IV. The area of the circle of radius x is

$$\odot x = 4\pi k^2 \operatorname{sh}^2\frac{x}{2k}.$$

V. The area of the plane region (extending to infinity) enclosed by a paracycle arc of length r and the axes starting from the endpoints of this arc equals

$$rk.$$

The volume of the space region (extending to infinity) enclosed by a parasphere region of area p and the axes starting from its boundary equals

$$\frac{1}{2} pk.$$

VI. It is a theorem of absolute geometry that the area of the spherical cap is equal to the area of the circle the radius of which equals the (straight) distance from the centre of the cap to the rim. The area of the sphere of radius x is

$$4\pi k^2 \operatorname{sh}^2 \frac{x}{k}.$$

VII. The volume of the ball of radius x is given by the expression

$$\pi k^2 \left[k \operatorname{sh} \frac{2x}{k} - 2x\right].$$

If the arc of the distance line to distance q corresponding to a segment of length p of the base line rotates about the base line, then the surface of revolution so obtained has area

$$\pi k p \operatorname{sh} \frac{2q}{k},$$

while the solid of revolution enclosed by this surface and its boundary discs has volume

$$\pi k^2 p \operatorname{sh}^2 \frac{q}{k}.$$

We note that the following two statements are direct consequences of I and II. The direction number of the curve $y=f(x)$ is

$$\operatorname{tg} \alpha = \frac{y'}{\operatorname{ch} \frac{y}{k}}.$$

The arc length of the curve $y=f(x)$ above the interval (a, b) is

$$s = \int_a^b \sqrt{\operatorname{ch}^2 \frac{y}{k} + y'^2}\, dx,$$

that is, for the arc length s of the curve given in parametric form $x(t)$, $y(t)$ we have

$$\left(\frac{ds}{dt}\right)^2 = \operatorname{ch}^2 \frac{y}{k} \left(\frac{dx}{dt}\right)^2 + \left(\frac{dy}{dt}\right)^2.$$

In the limit $k \to \infty$ each of the theorems above turns into the corresponding theorem of Euclidean geometry.

The treatment in this section does not comply with the modern standards of mathematical rigour. The reader should not forget, however, that in 1831 the exact concepts of differential geometry (continuity arguments, arc length, surface area, etc.) were unknown. Their exact treatment, even for Euclidean space, evolved only at the end of the century. JÁNOS BOLYAI gives the most rigorous treatment possible, also as compared to the greatest mathematicians of his time. This section, however, is still more concise than the others; namely it includes an abundance of facts reaching far beyond the fundamentals of the new geometry. The results convince us that, even if laconic in description of the proofs, BOLYAI thought over everything. Below, by expounding the determination of the direction number and the line element in more detail, we give an idea of how he might obtain his formulas.

1. We consider the arc $\widehat{PQ}$ of the curve $y=f(x)$. $\overline{PM}$ and $\overline{QN}$ are ordinates; the coordinates of P are denoted by x and y. The hypercycle t that corresponds to the x-axis and the distance y intersects QN at U. For the triangle PQU by the sinc law (see Fig. 27)

$$\frac{\sin \varphi}{\sin \psi} = \frac{\operatorname{sh}\dfrac{\Delta y}{k}}{\operatorname{sh}\dfrac{\Delta \tau}{k}}.$$

If Q approaches P, then $\varphi \to \alpha$ and $\psi \to R-\alpha$. Therefore, passing to the limit in our equation,

$$\operatorname{tg} \alpha = \lim_{\Delta x \to 0} \frac{\operatorname{sh}\dfrac{\Delta y}{k}}{\operatorname{sh}\dfrac{\Delta \tau}{k}} = \lim_{\Delta x \to 0} \frac{\Delta y}{\Delta \tau}.$$

Hence the assertion follows since by the remark to §30 $\dfrac{\Delta t}{\Delta \tau} \to 1.$

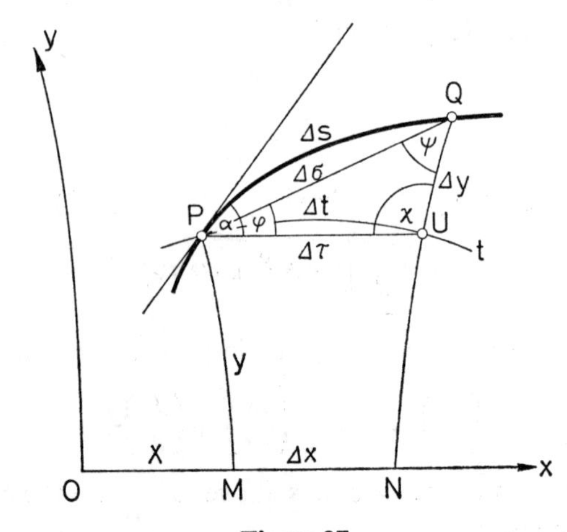

Figure 27

2. For the triangle PQU by the sine law

$$\frac{\sin\varphi}{\sin x} = \frac{\operatorname{sh}\frac{\Delta y}{k}}{\operatorname{sh}\frac{\Delta\sigma}{k}}.$$

If Q approaches P, then $\varphi \to \alpha$ and $x \to R$. Thus, in the limit, our equation yields

$$\sin\alpha = \lim_{\Delta x\to 0}\frac{\operatorname{sh}\frac{\Delta y}{k}}{\operatorname{sh}\frac{\Delta\sigma}{k}} = \lim_{\Delta x\to 0}\frac{\Delta y}{\Delta\sigma}.$$

Similarly, using the relation $\frac{\Delta\tau}{\Delta t}\to 1$ mentioned above,

$$\cos\alpha = \lim_{\Delta x\to 0}\frac{\Delta\tau}{\Delta\sigma} = \lim_{\Delta x\to 0}\frac{\Delta t}{\Delta\sigma},$$

and by addition of the squares of these two equations

$$\lim_{\Delta x\to 0}\frac{(\Delta y)^2+(\Delta t)^2}{(\Delta\sigma)^2} = 1.$$

This gives the assertion, since $\frac{\Delta s}{\Delta\sigma}\to 1$. The latter relation holds for curves with continuous direction angle (smooth curves); its proof will not be presented here.

For completing the discussions about area, volume and surface area to become perfect from the viewpoint of mathematical rigour, one should first define these very concepts in an exact fashion and then work out the material of the section based on the exact definitions. All this, however, is beyond the scope of the present remarks.

On the other hand we remark — not striving for a strict formulation this time — that hyperbolic space may be considered Euclidean in the small. That is, if we perform calculations within a small neighbourhood according to the rules of Euclidean geometry, then the results obtained have small relative error. As the neighbourhood decreases indefinitely, the relative error tends to 0. This statement is related to the fact that in the limit $k\to\infty$ the theorems of hyperbolic geometry turn into the Euclidean ones. For instance, the fraction $\frac{a}{k}$ which appears in trigonometry tends to 0 for either decreasing a or increasing k. If we accept that hyperbolic geometry is "Euclidean in the small" then, point by point, also the considerations of the present section of the *Appendix* become exact.

§33

According to this section it is an open question which of the systems S and Σ is valid in reality. Further, if the valid system were S, it is unknown what value the parameter k has in that S describing the real world.

In the latter case it would suffice to know the value H belonging to one single distance h, and k could be determined by the relation

$$k = h:\ln H.$$

In System S, the parameter k is the distance to which the corresponding K is equal to e. This distance can serve as a *natural unit* of distance-measuring.

The difference between Euclid's and BOLYAI's geometries is indicated by the existence of a natural unit. In Euclidean geometry there is no natural unit, and only after *arbitrarily* selecting a unit distance can we assign a measure to any distance.

LOBACHEVSKY set himself the task of deciding (indirectly) by measurement whether Euclidean geometry is the geometry valid in the real world. His efforts brought no results. As the expected value of the parameter k is very large, instead of terrestrial measurement he tried to reach his aim by way of astronomy, using configurations composed of distances which are immense even as compared with the diameter of the orbit of the Earth.

We give the following simplified account of LOBACHEVSKY's procedure. Suppose that points A and B are two positions of the Earth moving from B to A in half a year. Let the rays of light coming from the fixed star C form angles R and $R-\alpha$ with the diameter $AB=a$ of the orbit of the Earth, and let (Fig. 28)

$$\overrightarrow{AM} \perp AB.$$

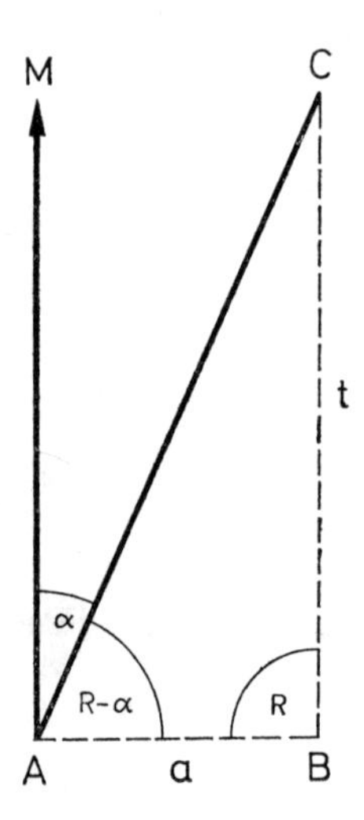

Figure 28

If S is the system which corresponds to reality, then

$$\Pi(a) > R - \alpha$$

and

$$k > \frac{a}{\operatorname{tg} \alpha}.$$

Lobachevsky carried out the measurement and found that

$$k > 166\,000 \cdot a.$$

JÁNOS BOLYAI had no confidence in LOBACHEVSKY'S procedure. Taking into account the practical difficulty of executing the measurement, he did not expect from contemporary measuring instruments that a decisive result would arise. In an ingenious remark, however, criticizing Lobachevsky's attempts he referred to the possibility of another measuring process. Namely he assumed that the gravitational force of attraction between two bodies is inversely proportional with the area of the sphere having the distance of the bodies for radius. Now the ratio of the areas of two spheres with radii a and b is

$$a^2 : b^2 \quad \text{in System } \Sigma,$$

$$\operatorname{sh}^2 \frac{a}{k} : \operatorname{sh}^2 \frac{b}{k} \quad \text{in System S.}$$

Therefore, in the motion of celestial bodies, data expected when assuming $a^2:b^2$ and data really observed may show a discrepancy on the basis of which the validity of System S can be concluded. But JÁNOS BOLYAI, isolated from the world, cut off even from the most primitive possibilities of scientific research, could not make measurements. He had to content himself with developing the new geometry and producing the synthesis of absolute geometry.

The vain efforts of LOBACHEVSKY had a fertilizing influence on subsequent researchers. The ingenuity of the above idea of BOLYAI has also been verified, to a certain extent, by the physics of later years. By their revolutionary courage to alter scientific views, the two great geometricians have also contributed to the radical change of physical concepts. When constructing the general theory of relativity, EINSTEIN relied on the work of BOLYAI and LOBACHEVSKY. *No doubt, the wonderfully rich flowering of the mathematical and physical concepts of space has sprung from the fundamental research of Bolyai and Lobachevsky.*

We already know that the structure of physical space is non-Euclidean. It is still more complicated than the space of classical non-Euclidean geometry. The parameter that characterizes the space is not constant, but varies from point to point depending on the mass distribution of matter filling the space.

Also this section of the *Appendix* is a clear evidence that JÁNOS BOLYAI regards physical space as an objective reality, geometry as a reflection of objective reality, and reality as accessible to cognition. Thus his scientific ideology is essentially that of dialectical materialism. Through this way of looking at science was he able to clear up the problem in labour for two thousand years.

§34

This section treats the question how the angle of parallelism corresponding to a given distance can be constructed. Before we begin to investigate the construction we have to define what is to be meant by a construction.

In plane geometry, by a compass construction we mean that we use only a compass with which we perform the following operations (steps):

1. Taking down the distance of two points by compass.
2. Drawing a circle (or circular arc) about a given point with a given span of the compass (radius).
3. Two circles being drawn, marking their points of intersection.

In a compass construction a finite number of steps 1, 2, 3 in any combination is allowed.

In plane geometry, by a straightedge construction we mean the application of a straightedge (ruler) in a finite number of steps according to the following prescription:

4. Laying the straightedge so as to be incident with two points simultaneously.
5. Drawing a line along the straightedge.
6. Two straight lines being drawn, marking their point of intersection.

By a compass-straightedge construction, or Euclidean construction, we mean any combination of the steps 1–6 and the following step 7, each performed only a finite number of times.

7. A circle and a straight line being drawn, marking their points of intersection.

If a construction can be carried out and leads to a correct result in both systems Σ and S, then it is called an *absolute construction.*

That construction of the regular hexagon which is based on the theorem "the circular arc corresponding to a chord equal to the radius is one sixth of the whole circle" is not absolute, it is valid in System Σ only. In System S, the side of the inscribed equilateral hexagon of the circle is greater than the radius of the circle.

On the other hand, the ordinary compass construction of the perpendicular through a given point to a given line is absolute. The usual construction of the perpendicular bisector of a segment is also an absolute construction.

The construction of the angle of parallelism described in the *Appendix* is not absolute because it can be applied only in System S. For, in System Σ, the circle with centre A and radius equal to ED is only tangent, and the rules for use of the instruments do not allow marking the point of contact. To be sure, in System Σ there is another, simpler construction of the line parallel to a given one.

§35

In this section the question is investigated how to obtain in the hyperbolic plane (that is, in System **S**) the distance corresponding to a given angle of parallelism by a compass-straightedge construction.

§36

This section deals with the compass-straightedge construction of the intersection point of a straight line and a plane, and also of the line of intersection of two planes.

§37

This section treats the construction of the corresponding point by a stereometric method.

§5 of the *Remarks* suggests a simple planar construction.

Given $\overrightarrow{AM} \parallel \overrightarrow{BN}$, *find the point* A' *that corresponds to* A *on the line* BN (Fig. 29).

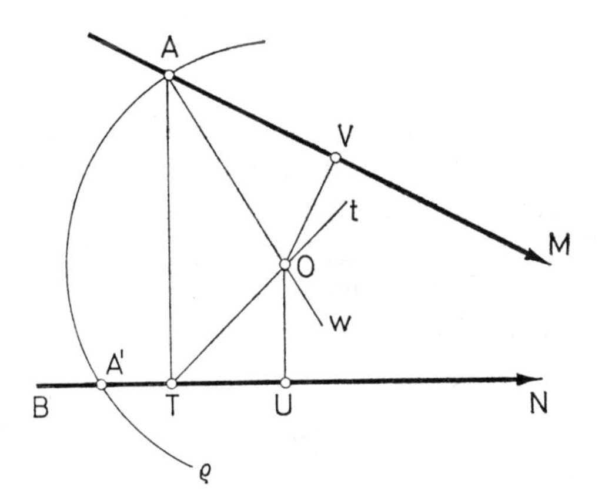

Figure 29

Let $\sphericalangle ATN = R$; denote its bisector by t, the bisector of $\sphericalangle MAT$ by w, and the common point of w and t by O. That point of intersection of BT and the circle ϱ with centre O and radius OA which belongs to the half-line complementary to $\overrightarrow{TN}$ is just the required A'.

It is easy to observe by what kind of ideas BOLYAI was guided when writing the last sections of the *Appendix*. On the parasphere, Euclidean geometry is valid if the role of straight lines is taken by paracycles. From the Euclidean theorems valid on the parasphere the theorems of absolute geometry and hyperbolic plane geometry can be derived in a simple way. Similarly, from the Euclidean constructions represented on the parasphere one can deduce absolute constructions. This is why we need to construct, using a compass and a straightedge, the points of the paracycle determined by certain data.

Thus, for instance, successively applying the constructions described in **§34** and **§37** we obtain a compass-straightedge procedure for solution of the following problem: given a point of an **L**-line and the line to an axis of this **L**-line, to construct the point of intersection of the line and the **L**-line. On the other hand, if we are given a point P and the axis $\overrightarrow{PM}$ of the **L**-line, further a line PX, and we have to construct the other point of intersection Q of PX and **L**, then the mid-point of PQ can be constructed by reduction to the case described in **§35** and Q is obtained by a reflection.

§38

This section treats the construction of the distances x which correspond to the values $X=2$ and $X=e$. In the latter case, however, only an approximative construction of the respective x, with any precision, is referred to.

§39

The aim of this section is to prove a theorem which is needed in the next sections. Let **T** and **T**′ be a pair of distance lines situated symmetrically with respect to the base line a. Let A, B lie on **T**, and C lie on **T**′. If C moves along **T**′ while A and B are fixed, then both the area and the angle sum of the partly, maybe, curvilinear triangle ABC (enclosed by two segments of straight lines and one segment of a distance line) remain unchanged; the angle sum is $2R$.

§40

This section deals with proving the theorem according to which any two triangles of equal area and with one side in common have equal angle sums.

§41

The theorem of the previous section can be generalized: *two triangles have equal areas if and only if they have equal angle sums.*

§42

According to this section, *in hyperbolic geometry the ratio of the areas of two triangles is equal to the ratio of the defects of the triangles.* (A similar theorem is true for spherical triangles, "defects" being replaced by "excesses".) This can also be expressed by the relation

$$\Delta = c\delta,$$

where δ and Δ denote defect and area, while c is constant for all triangles.

Next we present another proof of this theorem. This proof appears in a note of JÁNOS BOLYAI from about 1834. Independently, GAUSS gave a similar proof of the theorem.*

Let $\overrightarrow{AB}\|\overrightarrow{EF}$, $\overrightarrow{AC}\|\overrightarrow{FE}\|\overrightarrow{GL}$ and $\overrightarrow{LG}\|\overrightarrow{AD}$. Further let $HK=HA$ and $\overrightarrow{KM}\|\overrightarrow{HB}$. Then obviously $DAHG\equiv MKHB$, that is, for the areas we have (Fig. 30)

$$\Delta_4 = \Delta_3.$$

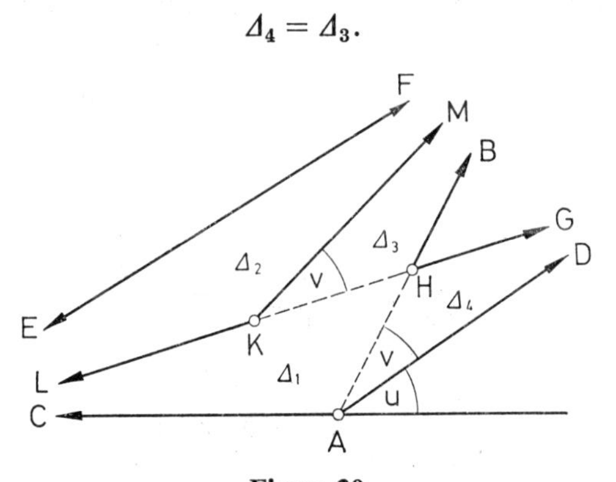

Figure 30

The figures Δ_1, Δ_3 and Δ_4 are limiting triangles, namely they can be obtained as limiting figures of triangles of variable shape where the area is fixed and the angles vary in such a way that one of them tends to 0. Also the figures Δ_2, $\Delta_2+\Delta_3$, $\Delta_1+\Delta_4$ and $\Delta_1+\Delta_2+\Delta_3$ are limiting triangles, but each of them has two angles equal to zero. Taking limits it can be proved that the theorem of **§41** remains valid for limiting triangles of this kind.

By **§41** the area of the triangle is a single-valued and — as it is easy to see — continuous function of the defect. Denote by $F(\delta)$ the area of the triangle of defect δ.

The lines in question cut the plane into pieces so that obviously

$$F(u+v) = \Delta_1+\Delta_2+\Delta_3 = \Delta_1+\Delta_2+\Delta_4 = (\Delta_1+\Delta_4)+\Delta_2 = F(u)+F(v).$$

The only continuous solution of this functional equation is

$$F(x) = cx.$$

The proof is complete.

* Cf. Part I, Section 4.

§43

This section treats one of BOLYAI's most beautiful theorems: the area and the defect of the triangle are related by the formula

$$\Delta = k^2\delta.$$

The notion of square in System S can be introduced in several ways. We must only take care that for $k \to \infty$ the quadrilateral satisfying the new definition should turn into the square defined according to System Σ. The only natural definition, however, seems to be the one used by BOLYAI: the square is a quadrilateral the four sides and four angles of which are equal. But in hyperbolic geometry the angles of the square are not right angles.

Let two straight segments $UX = VY$ passing through the point O be perpendicular bisectors of each other. Let the perpendiculars drawn through their endpoints intersect at A, B, C and D. The figure $ABCD$ is a square, $OX = p$ is its *parameter*. Obviously, if p is not greater than the distance of parallelism to the angle $\frac{R}{2}$, then there is a square of parameter p (Fig. 31).

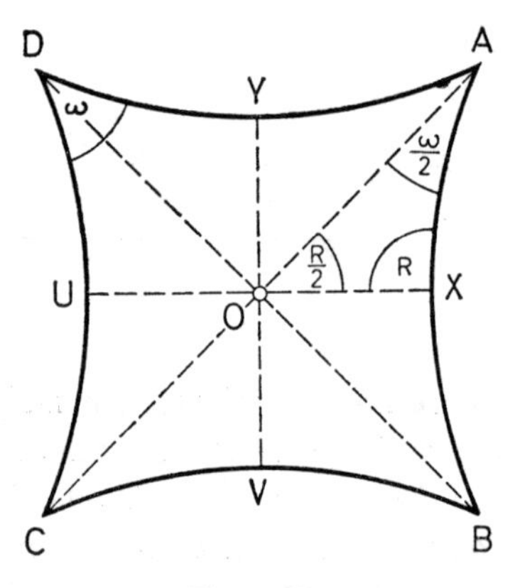

Figure 31

The area of the square can be dissected into eight figures congruent to the right triangle OXA. Since the area of one such triangle is

$$k^2\left(\frac{R}{2} - \frac{\omega}{2}\right),$$

where ω denotes the angle of the square, the *area of the square* is equal to

$$4(R-\omega)k^2.$$

We call *maximal limiting triangle* the limiting triangle of maximal area.

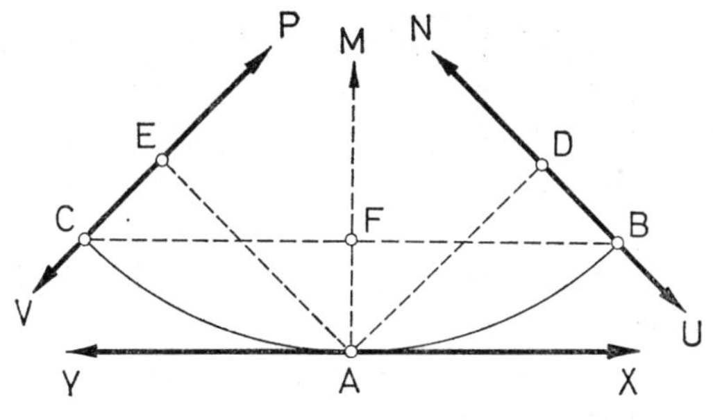

Figure 32

The triangle has maximal defect $2R$ when all sides are parallel. Since the maximal limiting triangle has defect π, its area is πk^2. Fig. 32 represents a maximal limiting triangle.

Let $\overrightarrow{AM} \perp \overrightarrow{XY}$, $\overrightarrow{NU} \| \overrightarrow{AX}$, $\overrightarrow{UN} \| \overrightarrow{AM} \| \overrightarrow{VP}$, $\overrightarrow{PV} \| \overrightarrow{XY}$, $\overrightarrow{AE} \perp \overrightarrow{PV}$, $\overrightarrow{AD} \perp \overrightarrow{NU}$, and let the paracycle arc $\widehat{BC}$ be contained in the paracycle of axis $\overrightarrow{AM}$ through A. If the paracycle arc $\widehat{AB}$ rotates about the axis $\overrightarrow{AM}$, then on the parasphere belonging to $\overrightarrow{AM}$ it *sweeps a disc (parasphere cap) of area equal to that of the maximal limiting triangle.*

Since AD is the bisector of $\sphericalangle MAX$, a symmetry argument yields

$$\sphericalangle FBN = \sphericalangle DAM = \frac{R}{2}.$$

Therefore, in view of our remarks to **§30,**

$$\widehat{AB} = k.$$

By **§21** it follows that the area of the surface swept by the arc $\widehat{AB}$ is πk^2, which completes the proof.

Next we show that there is a compass-straightedge construction of the square with area equal to that of the maximal limiting triangle.

By the foregoing, the angle ω of the square of area $k^2\pi$ satisfies the equation

$$\pi = 4(R-\omega);$$

thus $\omega = \frac{R}{2}$. Therefore, in case of the square of area πk^2, the angles of the right triangle OXA of Fig. 31 are $\frac{R}{2}$, R and $\frac{R}{4}$, respectively.

From the triangle OXA, according to **§31**, II,

$$\sin \Pi(p) = \sin \frac{R}{2} : \cos \frac{R}{4} = 2 \sin \frac{R}{4} = \sqrt{2-\sqrt{2}}.$$

If the unit distance is given, then $\sqrt{2-\sqrt{2}}$ can be constructed in the Euclidean plane by compass and straightedge. The same is true on the parasphere. Then, however, the angle whose sine is $\sqrt{2-\sqrt{2}}$ can also be constructed on the parasphere. Compass-straightedge constructions on the parasphere can be transformed into stereometric constructions and these, in turn, into compass-straightedge constructions in the plane. Thus the angle $\Pi(p)$ can be constructed. Hence p can also be constructed.

If p is available, then the corresponding square can be obtained by an absolute construction using compass and straightedge.

If we can also find a compass-straightedge construction for the radius of the circle the area of which is equal again to the area of the maximal limiting triangle, then together with the previous construction a solution of the problem of squaring the circle is obtained.

For constructing a circle of area $k^2\pi$ in the plane we first construct the angle $\frac{R}{2}$. Then we construct the distance of parallelism that corresponds to the angle $\frac{R}{2}$; erecting a perpendicular at one of its endpoints and laying the angle $\frac{R}{2}$ at the other endpoint we obtain the line $MFBN$ of Fig. 32. On the line FM we can construct the point A which corresponds to the point B of the line BN. The area of the circle of radius $\overline{AB}$ is $k^2\pi$, since by **§32**, VI, the area of this circle is equal to the surface area of the cap bounded by the circle that the paracycle arc $\widehat{AB}$ describes on the parasphere, and since we have already seen that the latter area equals $k^2\pi$.

Thus we have solved the problem of squaring the circle in the hyperbolic plane by a compass-straightedge construction. In fact, we have constructed a square and a circle of equal area coinciding with the area of the maximal limiting triangle.

PART IV

THE WORK OF BOLYAI AS REFLECTED BY SUBSEQUENT INVESTIGATIONS

In this final part of our book we sketch the spread, effect and further evolution of the discovery and ideas of JÁNOS BOLYAI after they had become known. A relatively complete discussion of this topic, even in concise form, would fill a thick volume. For this reason, and since we will not assume that the reader has had a comprehensive and thorough mathematical education, we only try to illuminate the renewing effect of BOLYAI's ideas on the whole of mathematics by outlining a few subjects.

In our exposition, we dwell on neither B. RIEMANN's fundamental paper *Über die Hypothesen, welche der Geometrie zugrunde liegen* (Leipzig, 1876) nor the formation and evolution of Riemannian geometries, since we deal with the effect of BOLYAI's thoughts rather than with their relations to other ideas. The interested reader, however, will be well oriented by G. VRANCEANU's masterly survey *Riemannian geometry* which has appeared as a chapter (pp. 227–256) of the book *The life and Work of János Bolyai* (in Hungarian; Bucharest, 1953, State Scientific Publishers).

Also recent works on the foundation of Euclidean geometry are not discussed in the sequel. At this point we content ourselves by referring to the excellent book *The Elementary Construction of Euclidean Geometry* by B. KERÉKJÁRTÓ (in Hungarian, 1937; French translation 1955).

BOLYAI's discovery was not only the first step leading to modern mathematics, his thoughts influenced the formation of modern ideas in natural science as well. The latter subject, however, must be left to a more competent author.

CHAPTER I

THE CONSTRUCTION OF GEOMETRY BY ELEMENTARY METHODS

1. FURTHER INVESTIGATIONS OF JÁNOS BOLYAI IN THE FIELD OF ABSOLUTE GEOMETRY

It turns out from the manuscripts of János Bolyai that he planned a perfect construction and detailed elaboration of geometry on a firm basis. He also planned a derivation of plane trigonometry making no use of stereometric relations. He struggled with the problem as to whether hyperbolic geometry is a system without inherent contradictions.

He was aware that even if plane hyperbolic geometry is a consistent system, the consistency of hyperbolic space does not follow. He established that in hyperbolic geometry there are three and only three kinds of "superficies undique uniformis". On the sphere the spherical geometry, on the parasphere the Euclidean geometry, and on the hypersphere the plane geometry of System **S** are valid. He objected to the loose and superficial treatment applied by his contemporaries to the problems of arc length, area and volume in the case of curved lines and surfaces. Though these questions did not exhaust his field of interest, his way of looking at problems shows that he always tried to penetrate into the very depth of geometry. He was ahead of his time, and several problems raised by him were solved by other mathematicians after decades only.

Below we call attention to some of his longer writings.

*The cubage of the tetrahedron.** Notes concerning the determination of the volume of the tetrahedron in hyperbolic space, made on several sheets and separate scraps of paper on different occasions.

*Responsio*** (from 1837). Competition paper submitted to the Jablonowskian Society. A sketch of his theory of imaginary quantities. We mention it here, because in §9 of that work BOLYAI discusses the coincidence between geometry valid on the sphere of imaginary radius and geometry of the hyperbolic plane. On a separate sheet attached to the draft of the *Responsio* he deals in more detail with defining the sphere

* Published by P. STÄCKEL in his book *The Geometrical Investigations of Farkas Bolyai and János Bolyai* (in Hungarian; Budapest, 1914, Hungarian Academy of Sciences; Part I, pp. 106—115).

** P. STÄCKEL: *The Geometrical Investigations of Farkas Bolyai and János Bolyai* (in Hungarian; Budapest, 1914, Hungarian Academy of Sciences; Part II, pp. 237–249).

of imaginary radius and with the unified formulation of trigonometric theorems for geometries valid on the three kinds of "everywhere uniform" surfaces in hyperbolic space.

*Comments** (from 1851). Criticism and analysis of LOBACHEVSKY's *Geometrische Untersuchungen*. Written in Hungarian.

*Raumlehre*** (from 1855). The first 75 sections of a manuscript left unfinished. Its purpose is to lay exact and complete foundations for geometry.

2. ELLIPTIC GEOMETRY

In connection with Riemann's paper cited above, the following question arose: In the construction of geometry, what is the role of those axioms which describe the straight line as an open curve? Which axioms are compatible with the assumption that *the straight line is a closed curve* — like the circle?

Investigations in this direction have led to the conclusion that it is possible to set up an incomplete system of axioms which remains consistent and becomes complete if one of the following axioms is added:

A. *There is a line e_0 and a point P_0, not in e_0, such that in the plane $[e_0 P_0]$ each of the lines passing through P_0 intersects e_0. (Axiom of intersection.)*

B. *There is a line e_0 and a point P_0, not in e_0, such that in the plane $[e_0 P_0]$ one and only one of the lines passing through P_0 does not intersect e_0. (Axiom of parallelism.)*

C. *There is a line e_0 and a point P_0, not in e_0, such that in the plane $[e_0 P_0]$ several of the lines passing through P_0 do not intersect e_0. (Axiom of non-intersection.)*

By adding **B** or **C**, respectively, to the incomplete system we obtain the axiom systems which characterize Euclidean and hyperbolic geometries. Addition of Axiom **A** provides the system of axioms for a third kind of geometry, the so-called *elliptic geometry*.

Thus the system of axioms for an *absolute geometry* can be set up so that by adding to it the axioms **A**, **B** or **C**, respectively, we obtain the system of axioms for elliptic, Euclidean, and hyperbolic geometries. (This absolute geometry, however, is not the same as that arising from Hilbert's system by cancellation of IV; its residual system of axioms is partly different.)***

* Ibidem, Part I, p. 127-158.

** Ibidem, Part II, pp. 251–288.

*** The reader can get acquainted with the elements of elliptic geometry from the book *Fragen der Elementargeometrie* by F. ENRIQUES (Leipzig–Berlin, 1923).

For the axiom system of elliptic geometry characterized by the property in question, see J. L. COOLIDGE: *The Elements of Non-Euclidean Geometry* (Oxford, 1909) and P. SZÁSZ: *The separation of elliptic, Euclidean, and hyperbolic geometries* (in Hungarian; Mat. Fiz. Lapok, Vol. 48, 1941).

We now mention a few theorems of elliptic geometry.

Two lines incident with a plane always intersect each other.

The line is a closed curve of finite length.

The angle sum of the triangle is $>2R$.

The areas of two triangles are to each other as are the excesses of the triangles.

On any line, for any point of the line there is a point at maximal distance from it.

The locus of points at maximal distance from a given point in the plane is a line called the polar of the given point. (The given point, in turn, is called the *pole* of this polar.)

The lines perpendicular to a given line in the plane meet at the pole of the given line.

We also mention that, on a spherical cap smaller than a hemisphere, if we regard great circles as straight lines, arc lengths of great circles as distances, and angles of intersection of great circles as angles, then from the theorems of spherical geometry we obtain theorems coinciding with theorems of elliptic geometry (as long as we remain in the interior of a circle of the elliptic plane). This will be discussed also later on.

3. THE COMMENTARY LITERATURE

The significance of a classical work is reflected by the wealth of the commentary literature on it. Even the list of the commentary literature related directly or indirectly to the *Appendix* would fill a separate volume.

JÁNOS BOLYAI, in later notes, made several remarks to the *Appendix*. He referred to a more suitable regrouping of the material, gave new proofs of some theorems, formulated new theorems, etc. Subsequently, more and more exacting mathematicians found the gaps of the treatment in the *Appendix* and gave simpler or different proofs; by providing the necessary details and suggesting a more instructive arrangement, they made the sketchy and condensed exposition more readable, adjusted it to the new points of view which had emerged as a result of the progress of mathematics, refined and enriched the methods.

We pick out a few examples.

The first complete adaptation in the commentary literature — FRISCHAUF's *Absolute Geometrie nach Johann Bolyai* (Leipzig, 1872, 96 pp.) written in text-book style — is distinguished by a convenient regrouping of the material, additions, and some simplifications.

In Vol. 12 (1903) of the journal *Mathematikai és Physikai Lapok,* as many as four authors have papers devoted to the *Appendix* (M. BEKE: *Bolyai's Trigonometry*; M. RÉTHY*: *Description of János Bolyai's "new, different world"*; J. KÜRSCHÁK: *On the*

* The name of MÓR RÉTHY deserves extra mention. He was the first Hungarian university professor to give a systematic exposition of absolute geometry in his course (1874). It was he who said that in our country, which apart from the two Bolyai's had had no significant mathematicians up to that time, all further scientific efforts ought to be rooted in the achievements of those two men.

Angle of Parallelism; P. SZABÓ: *On a Basic Theorem of Absolute Geometry*; all in Hungarian). Moreover, included in that volume is L. SCHLESINGER's memorial speech *János Bolyai* (in Hungarian) which, because of certain passages, can also be ranked among the commentary literature of the *Appendix.*

The book *Die nichteuklidische Geometrie* by BONOLA and LIEBMANN (*Wissenschaft und Hypothese,* IV, 1908), should be mentioned especially for the reason that it provides a detailed description of the history of non-Euclidean geometry.

We have already referred to F. ENRIQUES' *Fragen der Elementargeometrie.* Its last section has been written by BONOLA. In the exposition of absolute and hyperbolic geometries it follows, at some places, a shorter and simpler way than Bolyai. In some important respects, it supplements the subject matter of the *Appendix.*

P. STÄCKEL's book *The Geometrical Investigations of Farkas Bolyai and János Bolyai* (in Hungarian; Budapest, 1914, Hungarian Academy of Sciences) in particularly interesting since, among other things, it contains JÁNOS BOLYAI's own remarks and addenda to the *Appendix.*

J. KÜRSCHÁK's paper *Constructions that are independent of the axiom of parallelism* (in Hungarian; Mat. Fiz. Lapok, Vol. 37, 1930), before dealing with the constructions, gives a 13 pages long treatment of the basic theorems of plane absolute geometry relying on Hilbert's system of axioms.

V. F. KAGAN has translated the *Appendix* into Russian and added abundant comments to it*. The translation was published in 1950 and fills 236 pages (about 100 pages of which are demanded by explanations and addenda arranged in footnotes and separate sections).

And now we select from the commentary literature three fairly comprehensive subjects which have been treated by several mathematicians: keeping together and separating the three geometries, independent construction of plane hyperbolic geometry without recourse to stereometry, and limiting the role of continuity. BOLYAI, LOBACHEVSKY and RIEMANN had already called attention to these subjects. In the *Appendix*, BOLYAI first develops a few of the basic theorems of absolute geometry, and it is only thereafter that, in connection with some theorems, he separates Euclidean and hyperbolic geometries from each other. We know from his notes that he has raised the question: "... to deduce plane trigonometry restricting ourselves to the plane...". Comparing LOBACHEVSKY's works with the *Appendix* it is striking that BOLYAI — in an effort to give short derivations of the theorems — often uses a continuity argument, while LOBACHEVSKY strives to avoid it.

Keeping the three kinds of geometry together and separating them when necessary: this method sheds light on their intrinsic relationship.

* JÁNOS BOLYAI: *Appendix* (in Russian; Moscow—Leningrad, 1950, Gostizdat).

4. FOUNDATION OF HYPERBOLIC PLANE GEOMETRY WITHOUT USING THE AXIOMS OF CONTINUITY

In the first decade of the twentieth century, as a result of HILBERT's previous investigations, mathematical research became very lively in connection with problems as which axioms are needed for the construction of Euclidean and non-Euclidean geometries, which axioms can be replaced by weaker ones, how the various geometries can most simply be developed from the respective systems of axioms. It is perhaps no exaggeration to say that the most permanent effect on the evolution of these subjects has been made by HILBERT's *Neue Begründung der Bolyai–Lobatschefskyschen Geometrie* (Math. Ann., Vol. 57, 1903) and HJELMSLEV's *Neue Begründung der ebenen Geometrie* (Math. Ann., Vol. 64, 1910).

To the axioms of incidence, order and congruence, Hilbert adds an axiom of parallelism suitable for hyperbolic geometry. Reduced to $3+4+5+1=13$ axioms, his system of axioms serves only for founding the concept of plane — in other words, for forming a concept of the plane which is independent of possible embedding in space. He shows that the foundations of hyperbolic geometry can be laid *without making any use of the axioms of continuity.*

HJELMSLEV shows that plane geometry can be built up also without the axioms of continuity or any assumption on the intersection or non-intersection of lines.

The influence of both articles can be traced in a richly flourishing literature the ripest fruits of which appear to be P. SZÁSZ' papers and F. BACHMANN's book *Aufbau der Geometrie aus dem Spiegelungsbegriff* (Springer-Verlag, 1959).

The basic idea of HJELMSLEV's construction is that congruences arise as products of reflections in lines, and also that the notion of a *pencil of lines* can be defined. The notion of a pencil of lines relies on the notion of isogonal correspondence introduced by BOLYAI. By a pencil of lines we mean a collection of lines in the plane which satisfies the following conditions:

1. To any point of the plane there is at least one line in the collection that passes through the point.

2. Isogonal correspondence on the lines of the collection gives rise to a *transitive* relation between points.

If S_l denotes reflection of the plane in line l belonging to the plane and 1 denotes the identity, then the necessary and sufficient condition for the lines a, b, c of the plane to belong to a pencil is expressed by the relation $(S_a S_b S_c)^2=1$. We do not deal here with Hjelmslev's considerations any longer; the interested reader may consult the excellent book of BACHMANN.

Next we give a more detailed account of Hilbert's above-mentioned work and its further development in papers of P. SZÁSZ.

To every line, keeping in view the concept of parallelism, Hilbert assigns two *ends* (points at infinity). He defines an *"end calculus"* based on reflections in lines, and introduces an analytic geometry with ends as coordinates. In this analytic geometry, *concurrent lines are expressed by a linear equation.* After the proof of this fundamental theorem, the celebrated paper consists of merely two further sentences:

"Having established that the equation of a point in line coordinates is linear it is easy to obtain PASCAL's special theorem involving a pair of lines and DESARGUES' theorem on triangles in perspective, as well as the remaining theorems of projective geometry. Then the well-known formulas of the BOLYAI–LOBACHEVSKY geometry can also be deduced without difficulty, and so this geometry can be built up relying only on Axioms I—IV."

P. SZÁSZ has raised the question whether that portion of the argument which proceeds through projective geometry can be eliminated and the treatment followed in the former part carried on smoothly in some other way. Also the affirmative answer to this question is due to him and found in those of his papers which deal with constructing the analytic geometry of the hyperbolic plane on the basis of HILBERT's end calculus (in one of these papers, hyperbolic trigonometry is deduced from analytic geometry). Next we sketch the substantial part of his results. For a more detailed account, see P. SZÁSZ: *Unmittelbare Einführung Weierstracher homogenen Koordinaten in der hyperbolischen Ebene auf Grund der Hilbertschen Endenrechnung* (Acta Math. Acad. Sci. Hung., Vol. 9, 1958, pp. 1–28).

Instead of HILBERT's axiom of parallelism, SZÁSZ introduces the following two axioms.

IV. 1. Let P and Q be two different points of the plane, and QY a half-line on one side of the line PQ. Then on the same side of PQ one can always find a half-line PX such that PX does not, but every half-line PZ interior to $\sphericalangle QPX$ does, intersect QY.

IV. 2. In the plane, a line e_0 and a point P_0, not incident with e_0, can be found so that there are at least two lines incident with P_0 which do not intersect e_0.

Thus SZÁSZ replaces Hilbert's system of axioms by a system consisting of $3+4+5+2=14$ axioms.

Using the concept of parallelism (considered in the hyperbolic plane), SZÁSZ introduces the concept of ends; every line has two ends. In defining the end calculus he follows Hilbert but, in order to state the definition of product in a simple form, he introduces certain functions. We give a more detailed account (including illustrations) of this fundamental concept-building of the paper (Figures 1–2).

The role of the end ε reminds us of the role of the positive unit element in a product which justifies the notation $E(O)=1$; obviously $E(t)\cdot E(-t)=1$.

The possibility of division can also be seen directly unless the divisor is the end σ. The existence and uniqueness of the mid-point of a segment can be derived from the three first groups of axioms. This permits the introduction of two more operations with ends: the operation $\frac{\sigma}{2}$ and, for a non-negative end σ, the operation $\sqrt{\sigma}$.

The mid-line of the strip bounded by the lines (σ, ω) and (O, ω) has ends ω and $\frac{\sigma}{2}$.

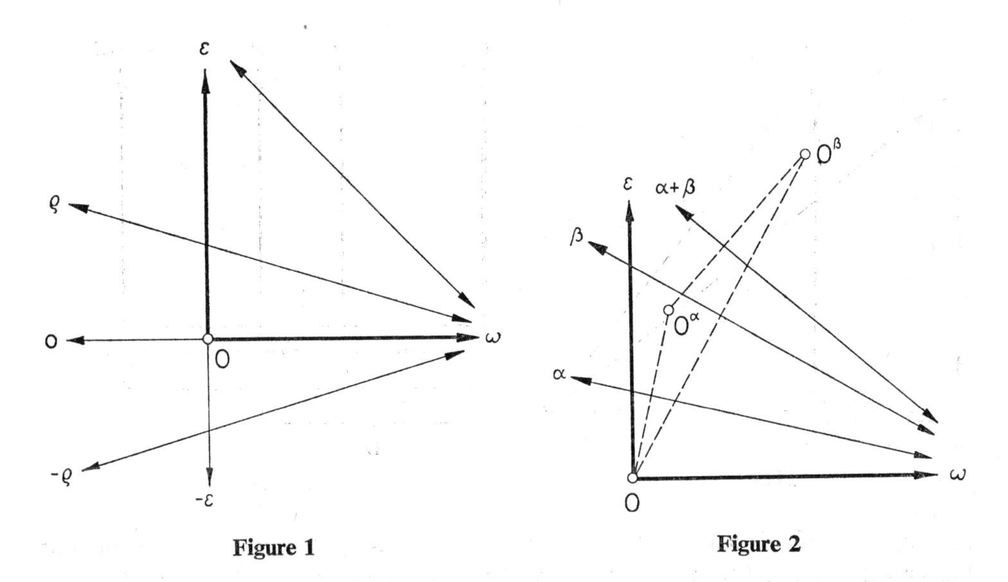

Figure 1

Figure 2

The square root of the positive end $\sigma = E(t)$ is defined by the relation $\sqrt{\sigma} = E\left(\frac{t}{2}\right)$.

Leaning especially on Group II of the axioms it becomes also possible to introduce the relation $\varrho < \sigma$ $(\sigma > \varrho)$ between ends. In the case of positive ends ϱ and σ, the relation $\varrho < \sigma$ means that the line (ϱ, ω) proceeds completely within the strip bounded by the lines (O, ω) and (σ, ω). In general, $\varrho < \sigma$ means that the end $\sigma - \varrho$ is positive.

The essence of the end calculus can be summarized in one single theorem:

The set of all ends different from ω *is an ordered field with respect to the two operations of the end calculus.*

Involved in the end calculus is a right angle $\omega O \varepsilon$ with vertex O. The other end of the lines $O\omega$ and $O\varepsilon$ is denoted by o and $-\varepsilon$, respectively. One of the half-planes determined by the line $O\omega$ contains the half-line $O\varepsilon$, while the other contains $O(-\varepsilon)$. The former of these half-planes contains all half-lines with positive ends and the latter contains those with negative ends; the strip formed by the lines (ϱ, ω) and $(-\varrho, \omega)$ with end ω has mid-line $O\omega$. Let α, β be two ends different from ω; further let O^α, O^β be the reflections of O in (α, ω) and (β, ω), respectively. The perpendicular bisector of the segment $O^\alpha O^\beta$ has ω for one end, the other end is called the *sum of the ends* α *and* β, and denoted by $\alpha + \beta$. The other end o of the line $O\omega$ plays the role of neutral element with respect to addition. The following rules are valid:

$$\alpha + o = \alpha, \quad \alpha + (-\alpha) = o, \quad \varepsilon + (-\varepsilon) = o,$$

$$\alpha + \beta = \beta + \alpha, \quad (\alpha + \beta) + \gamma = \alpha + (\beta + \gamma).$$

We take the distance from O to the points of the half-line $O\omega$ and to the points of

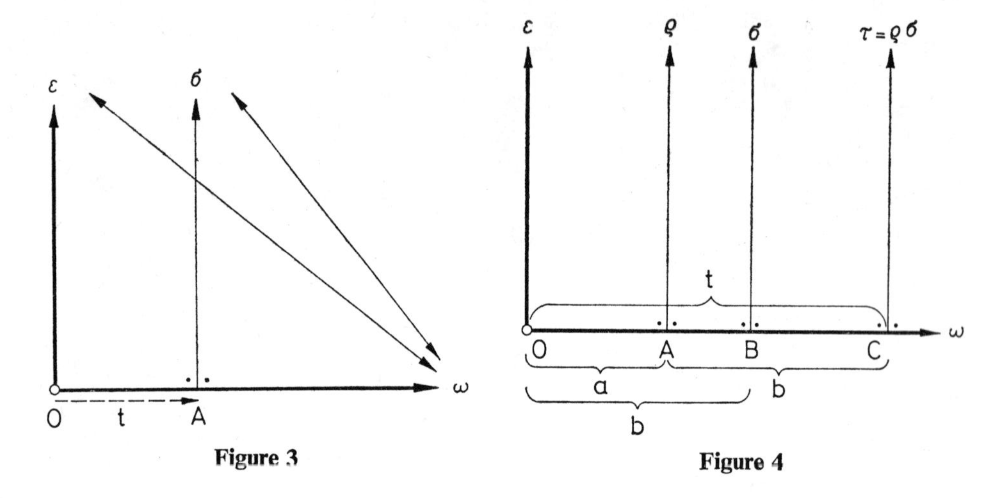

Figure 3

Figure 4

the extension of this half-line to be positive and negative, respectively. At point A of the signed segment OA, or segment t for short, we erect a perpendicular to $O\omega$. For the positive end σ of this perpendicular we write (Fig. 3)

$$\sigma = E(t).$$

Denote by ϱ and σ the positive and of the perpendicular to $O\omega$ at A and B, respectively, and by τ that of the perpendicular at the point C determined by the relation $OC=OA+OB$ or $t=a+b$. By the *product of the positive ends* $\varrho=E(a)$, $\sigma=E(b)$ we mean the end $\tau=E(t)=E(a+b)$; we write $\varrho\sigma=\tau$ or, equivalently, $E(a)\cdot E(b)=$ $=E(a+b)$ (Fig. 4).

The notion of product can be extended to o and negative ends: $(-\varrho)\sigma=\varrho(-\sigma)=$ $=-\varrho\sigma$, $(-\varrho)(-\sigma)=\varrho\sigma$, $\xi\cdot o=o\cdot\xi=o$.

The following rules are valid:

$$\varepsilon\varrho = \varrho\varepsilon = \varrho, \quad \varrho\sigma = \sigma\varrho, \quad (\pi\varrho)\sigma = \pi(\varrho\sigma),$$
$$\pi(\varrho+\sigma) = \pi\varrho+\pi\sigma.$$

With the help of the segment function $E(t)$, for the sake of having simple and concise formulations later on, Szász introduces three further segment functions:

$$C(t) = \frac{E(t)+E(-t)}{2}, \quad S(t) = \frac{E(t)-E(-t)}{2}, \quad T(t) = \frac{E(t)-E(-t)}{E(t)+E(-t)}.$$

They resemble the hyperbolic functions and, in particular, satisfy the relations

$$C(t)\cdot C(t)-S(t)\cdot S(t) = 1,$$
$$C(a+b) = C(a)\cdot C(b)+S(a)\cdot S(b),$$
$$S(a+b) = S(a)\cdot C(b)+S(b)\cdot C(a).$$

On the other hand, $E(t)$ resembles the exponential function.

The revised treatment of the end calculus given by Szász and described above makes possible the introduction of the "Weierstrass homogeneous coordinates" which represent the points and lines of the hyperbolic plane. This is prepared by the following theorem of SzÁsz:

There is a one-one correspondence between all points of the hyperbolic plane and those triples (ξ, η, ζ) *of ends different from* ω *which satisfy the conditions*

$$\xi^2-\eta^2-\zeta^2 = 1, \quad \xi > \sigma.$$

It can be implemented by assigning to the point given by the mixed pair of elements (t, λ) *the triple of ends*

$$\xi = C(t)+\frac{1}{2}\lambda^2\cdot E(-t),$$

$$\eta = \lambda\cdot E(-t),$$

$$\zeta = S(t)+\frac{1}{2}\lambda^2\cdot E(-t).$$

For any pair of points (ξ, η, ζ), (ξ', η', ζ') we have the relation

$$\xi\xi'-\eta\eta'-\zeta\zeta' < o.$$

SzÁsz introduces line coordinates for specifying a line which is *oriented* towards one of its ends. If the ends involved are different from ω, then the line with ends λ_1, λ_2 oriented towards λ_2 has the homogeneous triple of ends (α, β, γ) given by

$$\alpha = \frac{\lambda_1\lambda_2-1}{\lambda_2-\lambda_1}, \quad \beta = \frac{\lambda_1+\lambda_2}{\lambda_2-\lambda_1}, \quad \gamma = \frac{\lambda_1\lambda_2+1}{\lambda_2-\lambda_1},$$

where

$$\alpha^2+\beta^2-\gamma^2 = 1.$$

The coordinatization can be extended to the point $(t, \lambda)=(0, o)$ and the line with end ω. If $t=0$, $\lambda=o$, then

$$\xi = o, \quad \eta = o, \quad \zeta = 1.$$

If the line (λ, ω) is oriented towards the end ω, then

$$\alpha = \lambda, \quad \beta = 1, \quad \gamma = \lambda.$$

The purpose of specifying the point and the line by homogeneous triples of end is clearly exhibited by the following theorem: *Incidence of the line* (α, β, γ) *with the point* (ξ, η, ζ) *is expressed by the equation*

$$\alpha\zeta+\beta\eta-\gamma\xi = o$$

(if instead of "$=o$" we have "$\neq o$", then the point is not incident with the line).

In possession of these foundations, SzÁsz easily arrives at the coordinate description of plane motion and the (planar) transformation called overturning: their effect on ends is given by a transformation of the form

$$\xi' = \frac{\alpha\xi+\beta}{\gamma\xi+\delta}$$

with $\mu>0$ for motion and $\mu<0$ for overturning, where

$$\mu = \alpha\delta - \beta\gamma.$$

Although there are no similar triangles in the hyperbolic plane, the theorems of (hyperbolic) trigonometry can be deduced easily on the basis of this construction. An unbroken, beautiful, improved version of the foundations laid by HILBERT has thus evolved.*

Also M. RÉTHY's paper mentioned above is partly concerned with a topic of this kind. For instance, with the aid of some witty ideas, the formulas of the three kinds of trigonometry are deduced jointly. These are obtained in a form containing an undetermined parameter. Restrictions on the parameter lead to the separation of elliptic, Euclidean and hyperbolic cases.

In one of his papers, relying on axioms I. 1—I. 8 of Hilbert's axiom system, further the axioms that characterize the separation of four planes belonging to a pencil, and finally the axioms of congruence which arise from the preceding axioms — but not relying on continuity — P. SZÁSZ derives the following theorem.

"In the plane of a line e and a point P, not on e, either all lines through P intersect e, or exactly one of them does not intersect e, or several of them do not intersect e, depending on which is the case for a unique line e_0 and a unique point P_0, not on e_0."

Then SZÁSZ formulates three axioms which state, respectively, that all lines through P_0 intersect e_0 *(axiom of intersection)*; exactly one of them does not intersect e_0 *(axiom of parallelism)*; several of them do not intersect e_0 *(axiom of non-intersection)*. If we attach one of these axioms to those used before and add a continuity axiom of Dedekind type, we obtain a *complete* system of axioms for a certain geometry (elliptic, Euclidean or hyperbolic geometry, in correspondence with the three axioms listed above).**

* The publications of P. SZÁSZ convince us of the living influence of JÁNOS BOLYAI's geometric ideas. A part of these publications are listed in the author's paper *Paul Szász: On the Occasion of his 60th Birthday* (in Hungarian; Matematikai Lapok, Vol. 13, 1962, pp. 9–21).

Mention should be made of J. C. H. GERRETSEN's paper *Die Begründung der Trigonometrie in der hyperbolischen Ebene* (Proc. Acad. Wetensch., Amsterdam, Vol. 45, 1942, pp. 360–366, 479–483 and 559–566). Namely this was the first perfect and fully detailed treatment, relying on Hilbert's work, of the foundations of the Bolyai—Lobachevsky geometry.

** P. SZÁSZ: *The separation of elliptic, Euclidean, and hyperbolic geometries* (in Hungarian; Mat. Fiz. Lapok, Vol. 48, 1941).

Hjelmslev also proves the inducing theorem of the previous paragraph making no use of the axiom of continuity, but relying only on planar axioms.* Naturally, the exclusion of spatial axioms makes the proof more complicated.

The first place among the works that give an independent treatment of the foundations of hyperbolic plane geometry is deserved by HILBERT's *Neue Begründung der Bolyai–Lobatschefskyschen Geometrie*; it does not even rely on the axiom of continuity.

The first presentation of hyperbolic trigonometry by merely planimetric tools is due to LIEBMANN.**
Since then, the subject has been considered in a number of articles.

* J. HJELMSLEV: *Neue Begründung der ebenen Geometrie* (Math. Ann., Vol. 64, 1907). — Einleitung in die allgemeine Kongruenzlehre (Mathematisk—fysiske Meddelelser, Vol. 8, No. 11 and Vol. 10, No. 1, 1929).

** H. LIEBMANN: *Elementare Ableitung der nichteuklidischen Trigonometrie* (Leipziger Berichte, Vol. 59, 1907).

CHAPTER II

THE CONSISTENCY OF NON-EUCLIDEAN GEOMETRIES

5. ON THE PROOF OF THE CONSISTENCY

The problem of consistency first appeared in the following form: do the theorems of geometry deduced from the system of axioms by means of logic contradict experience concerning the real (physical) space? In the times when mathematicians in their reasoning usually identified experience about the real space with the Euclidean interpretation of this experience, even putting the correct question meant a progress. For instance, the following theorem was considered important: if absolute geometry is valid in reality, then by performing a perfectly accurate measurement on one concrete figure it can be decided whether the Euclidean system or the geometry corresponding to a well-defined and finite value of k holds in the real world. The accuracy of measurement, however, is always restricted, thus even in an empirical way we cannot decide anything else than which geometrical system is *suitable* for expressing the properties of physical space. Neither this is a permanent decision valid "forever", since "suitableness" depends on the state of progress made in understanding the physical world. Consequently, this question leads out of mathematics; we do not dwell on it any longer.

By the consistency of a geometrical system we mean also another requirement: the absence of intrinsic contradictions.* In other words, that from the system of axioms it is impossible to deduce two theorems one of which is the negation of the other. How can the consistency of an axiom system be proved? This is our next subject.

Let **H** denote the system of axioms for hyperbolic geometry. System **H** deals with *"points"*, *"lines"*, *"planes"* and those relations between them which we express by the words *"incident"*, *"between"*, *"congruent"*, *"parallel"*, *"continuous"*. We have put these words in quotation marks to distinguish them from the corresponding ordinary (Euclidean) notions. Take certain elements of Euclidean geometry — not necessarily points or basic elements — and regard them as *"points"*; take certain other kind of Euclidean elements for *"lines"*, and a third kind for *"planes"*. Consider certain rela-

* From BOLYAI's notes it turns out that he has raised the question: is System **S** (intrinsically) consistent? He discovered also that consistency of hyperbolic planimetry does not imply that of stereometry.

tions between these Euclidean notions — which may even be composite relations derived from the Euclidean basic relations — to mean *"incidence"*, *"betweenness"*, *"congruence"*, *"parallelism"*, *"continuity"*.

This procedure is similar to writing down the explanation of signs to a map (a small circle is a "town", a dotted line is the "boundary of a county", etc.). Thus, so to say, we obtain a map in Euclidean geometry to represent the logical structure of the axiom system **H**. On this map, the axioms of **H** are reflected in the form of some Euclidean assertion. If, for instance, we replace the basic elements and basic relations appearing in an axiom **P** of System **H** by those Euclidean elements and relations which are their counterparts according to the explanation of signs — and leave the meaning of everything else unchanged —, then the axiom turns into a Euclidean proposition **P′**. If this transformation converts every axiom of **H** into a correct Euclidean theorem, then the map represents the elements of the axiom system correctly. Such a map is called a *model*, and we say that *the model realizes the axiom system* **H**.

If System **H** *can be realized by a Euclidean model, then* **H** *is a consistent system of axioms.*

Indeed, the counterpart of a deduction in System **H** is a deduction concerning the model, and the counterpart of the final theorem is a Euclidean theorem. If two deductions lead to contradictory theorems in **H**, then the two corresponding deductions lead to contradictory theorems in the model. Since two deductions which start from theorems of Euclidean geometry cannot lead to contradictory theorems, also **H** must be consistent. This conclusion relies on the consistency of Euclidean geometry, which is supported by intuition but has not been proved mathematically. In this sense, by using a model we can establish only "relative consistency".

The procedure is called the *method of models*. It consists of the following steps. 1. Finding the correspondence with the model. 2. Checking that the axioms of System **H** turn into correct theorems in the model.

As an example, we give a simple model of elliptic geometry. In Section 2 of this chapter we already mentioned that elliptic plane geometry can be represented on a spherical cap smaller than a hemisphere as on a map. In this way, however, only a *part* — a circular domain — of the elliptic plane is represented on a *part* of the sphere. The *whole* elliptic plane cannot be represented by using the *whole* sphere. Indeed, the theorem — involving the whole of the elliptic plane — which says that two lines have *only* one point in common is not realized on the sphere since two great circles intersect each other at two antipodal points of the sphere. This difficulty can be removed by the logical "trick" of regarding a pair of antipodal points as a single point.

We thus obtain the following model.

The *"plane"* is represented by the sphere, a *"point"* by a pair of antipodal points of the sphere, a *"line"* by a great circle of the sphere, a *"line segment"* by an arc of a great circle (and the antipodal arc), an *"angular domain"* by a spherical lune (and the antipodal spherical lune); *"equal line segments"* are represented by arcs of great circles of equal length, and *"equal angles"* by equal spherical angles.

From the correspondence described above it can be seen easily that a circle of the model sphere represents a *"circle"* of the elliptic plane.

We mention that the concept of model and its application have first appeared in JÁNOS BOLYAI's work. In contrast to later times when the representation of the hyperbolic plane by a model served for proving relative consistency, BOLYAI had used his model as a tool for abridging the deduction of theorems in hyperbolic geometry. Indeed, BOLYAI knew that each surface in non-Euclidean space has its own peculiar intrinsic geometry and so, presumably, there exists a particular surface on which just the Euclidean geometry is realized. This surface is the parasphere.

Though compelled to conciseness, BOLYAI gave unmistakable instructions which pointed at the heart of the matter and indicated how the exacting reader can develop the subject in full. Also, BOLYAI has nearly discovered the significance of traversing the way in opposite direction. For this purpose the following question should have been raised: is there a particular surface in Euclidean space the intrinsic geometry of which coincides with the geometry of the hyperbolic plane?

This question was proposed and answered in the affirmative by BELTRAMI no less than 36 years later. BELTRAMI's discovery has served as a starting-point for the elaboration of various representations.

6. BELTRAMI'S MODEL

We have seen that a circular domain of the elliptic plane can be represented on a spherical cap smaller than the Euclidean hemisphere. Then the question naturally arises whether one can also find a map surface suitable for representing the hyperbolic plane, again in the sense that arc length of a line in the surface and angle of intersection of two lines in the surface should mean arc length and angle of intersection of the lines represented. A map of this kind is said to be distance- and angle-preserving.

Let the parametric expression of the surface in Cartesian coordinates be

$$x = x(u, v), \quad y = y(u, v), \quad z = z(u, v).$$

Let the family of the u and v parameter lines of the surface form an orthogonal lattice. Then for the arc length of the surface curve $u=u(t)$, $v=v(t)$ we have

$$\frac{ds}{dt} = \sqrt{E\left(\frac{du}{dt}\right)^2 + G\left(\frac{dv}{dt}\right)^2}.$$

Further, the Gaussian curvature of the surface is

$$K = -\frac{1}{\sqrt{EG}}\left\{\frac{\partial}{\partial u}\left(\frac{1}{\sqrt{E}}\frac{\partial\sqrt{G}}{\partial u}\right) + \frac{\partial}{\partial v}\left(\frac{1}{\sqrt{G}}\frac{\partial\sqrt{E}}{\partial v}\right)\right\}$$

(one should not forget that everything is being considered in Euclidean geometry).

If we now take a plane in which the Cartesian coordinates of a point are denoted by $(\bar{x}, \bar{y})$, then the relations

$$\bar{x} = u, \quad \bar{y} = v$$

define a one-to-one mapping between the Euclidean surface treated above and the plane being considered. We require that this mapping be distance- and angle-preserving. The requirement is satisfied if and only if for E, G at a point (u, v) of the surface and $\bar{E}$, $\bar{G}$ at the corresponding point $(\bar{x}, \bar{y})$ we have

$$E = \bar{E}, \quad G = \bar{G}.$$

That is, if for the arc length in the plane

$$\frac{d\bar{s}}{dt} = \sqrt{E\left(\frac{d\bar{x}}{dt}\right)^2 + G\left(\frac{d\bar{y}}{dt}\right)^2}.$$

In the hyperbolic plane

$$E = \mathrm{ch}^2 \frac{\bar{y}}{k}, \quad G = 1.$$

Thus the Gaussian curvature of the corresponding surface is

$$K = -\frac{1}{\mathrm{ch}\,\frac{\bar{y}}{k}} \frac{\partial^2}{\partial \bar{y}^2}\left(\mathrm{ch}\,\frac{\bar{y}}{k}\right) = -\frac{1}{k^2} < 0,$$

a constant negative value independent of the point (u, v) in the surface. A surface of this kind is said to have constant negative curvature.

For the shortest path between two points of the hyperbolic plane, its counterpart is the shortest path between two points in the surface (a *geodesic* of the surface).

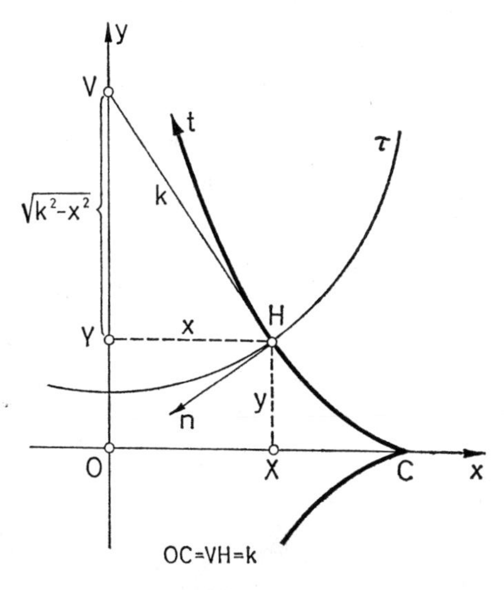

Figure 5

For the Euclidean and elliptic planes — by complete analogy with the previous argument — we obtain that $K=0$ and $K=\frac{1}{k^2}>0$ (a constant), respectively.

Surfaces satisfying the conditions $K>0$, $K=0$, $K<0$ are realizable. A simple calculation shows that, for instance, the sphere satisfies $K>0$, while the plane and the cylinder satisfy $K=0$. The condition $K<0$ is fulfilled, for instance, by the *pseudo-sphere*. We next describe how the pseudo-sphere can be generated.

The pseudo-sphere arises by rotation of the tractrix. The tractrix, in turn, can be obtained in the following way (Fig. 5).

There is a ship at the point H. A tug-boat V proceeds along the line y (the tug-boat tows the ship by a rope of length $VH=k$). In every moment the ship moves in the direction of the tight rope. Thus the ship describes a curved path the tangent of which has constant length k from the ship to the towing-path. From the right triangle YHV we immediately obtain the differential equation of the path of the ship, the tractrix, in the system of coordinates with axes $x \perp y$:

$$\frac{-\sqrt{k^2-x^2}}{x}=\frac{dy}{dx}.$$

The curve t could have been defined also by shifting a circle of radius k along the y-axis. The orthogonal trajectories of the pencil of circles so obtained are tractrices which can be shifted into each other along the y-axis.

If the curve t rotates about the y-axis and a circle of radius k rotates about one of its diameters, then they describe a pseudo-sphere and a sphere, respectively. Their Gaussian curvatures are easy to calculate:

$$K=\frac{1}{k^2} \quad \text{for the sphere,}$$

$$K=-\frac{1}{k^2} \quad \text{for the pseudo-sphere.}$$

To sum up our considerations, the planimetries of the three geometries *can be represented on surfaces of constant curvature,* according to the different cases, as follows:

$$\left.\begin{array}{l}\text{the elliptic}\\ \text{the Euclidean}\\ \text{the hyperbolic}\end{array}\right\} \text{ plane on a surface with } \begin{cases}K>0\\ K=0\\ K<0.\end{cases}$$

This representation (one-to-one correspondence) consists in associating the following pairs of notions:

in the plane	*on the surface*
point	point
straight line	geodesic
distance	geodesic arc length
angle (formed by two straight lines)	angle (formed by two geodesics).

The theorems of the geometry constructed axiomatically are realized on the appropriate map surface. Thus, we have proved the theorem according to which *if Euclidean geometry is consistent then so is hyperbolic planimetry.*

The first proof of the consistency of hyperbolic planimetry was given by BELTRAMI* in 1868. What he proved was precisely that hyperbolic geometry is realized on a portion of the pseudo-sphere.

In treating the representation of the elliptic and hyperbolic planes we have found only that a *part* of the elliptic plane can be represented on a *part* of the sphere, and a *part* of the hyperbolic plane can be represented on a *part* of the pseudo-sphere.

In 1901, HILBERT proved that there is *no* surface of constant curvature on which the whole hyperbolic plane or the whole elliptic plane could be represented in a one-to-one, distance- and angle-preserving manner.**

Therefore, with the help of the pseudo-sphere and the sphere respectively, by BELTRAMI's method we can only prove that hyperbolic and elliptic planimetries do not lead to a contradiction in a *part* of the plane. In the case of the sphere, by uniting antipodal points we obtain a model which establishes the consistency of elliptic planimetry in the *whole* elliptic plane.

We prove the consistency of hyperbolic planimetry in the whole hyperbolic plane and the consistency of hyperbolic stereometry by another model. This will be done in the next section.

7. THE CAYLEY—KLEIN MODEL

We prove that hyperbolic stereometry is consistent. The proof will be accomplished by the method of models. The model of hyperbolic space will be provided by a finite part of Euclidean space.

Let "*space*" mean the portion of Euclidean space interior to a given sphere **G** (so the points on, or exterior to .the sphere do not belong to the "space". Words in quotation marks refer to notions related to hyperbolic space.

"Point" shall mean a point interior to **G**. *"Line"* means a chord of **G** deprived of its endpoints. *"Plane"* means the part interior to **G** of a plane which intersects **G** (that is, the chord plane deprived of its circumference).

"Segment" means a segment both endpoints of which are interior to **G**. *"Half-line"* means the portion interior to **G** of a half-line which has its origin interior to **G**. *"Half-plane"* means the portion interior to **G** of a half-plane from the boundary line of which a chord is cut out by **G**. *"Angular domain"* means the portion interior to **G** of an angular domain which has its vertex interior to **G** (Fig. 6).

* BELTRAMI: *Saggio di interpretazione della geometria non-euclidea* (Giornale di Mathematiche, Vol. 6, 1868).

** Appears e.g. as Appendix V of D. HILBERT's *Grundlagen der Geometrie* (7th edition, Teubner, 1930).

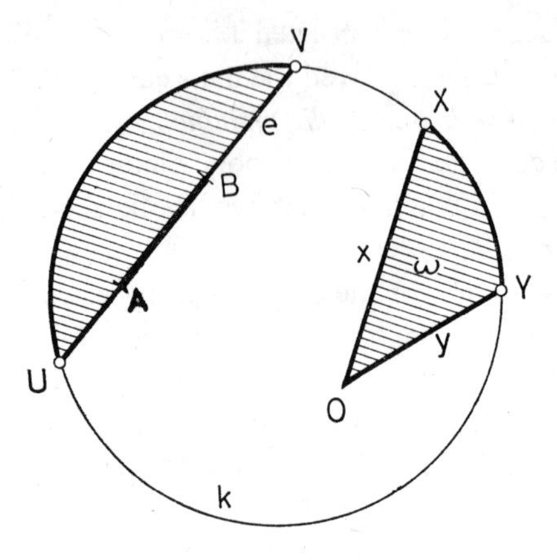

Figure 6

"Incidence" means incidence interior to the sphere. Also *"betweenness"* means betweenness, provided that each of the three points involved is interior to **G** (and, of course, they are on one and the same chord).

"Decreasing sequence of segments" will mean a decreasing sequence of segments even the first of which is interior to **G** (Fig. 7). *"Two parallel lines"* mean two chords with one endpoint in common (naturally, the endpoints of these chords do not belong to the *"lines"* either now).

The definition of *"congruence"* is more complicated. By the statement *"figure Γ_1 is congruent with figure Γ_2"* we mean that there is a collineation* (projective point

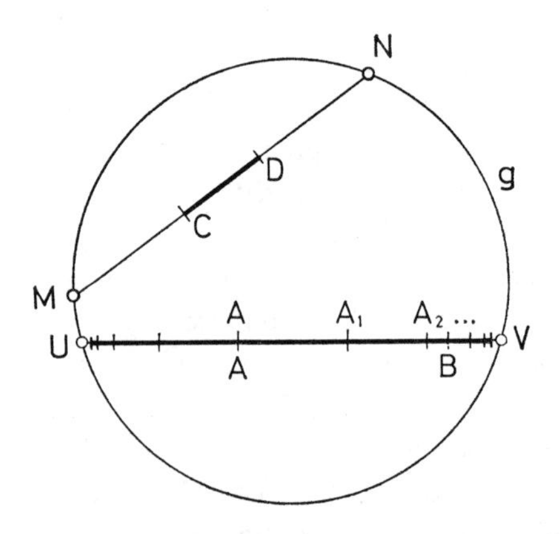

Figure 7

* A one-to-one point transformation is called a collineation if it takes the points of a line into points of a line (the ideal point being also admitted). Collineations preserve planes and cross ratios as well.

transformation) which takes **G** into **G** and Γ_1 into Γ_2 (Γ_1 and Γ_2 are assumed to be interior to **G**).

So, with hyperbolic space we have associated a Euclidean model. Next we must check whether the model realizes the axioms of hyperbolic geometry. Since this procedure involves the application of theorems on collineations that take the sphere into itself, we break the course of our considerations, present the theorems we need, and only then return to checking the axioms.

Theorem 1. *A collineation which maps a sphere onto another sphere takes all points interior to the sphere, and only them, into points interior to the other sphere.*

Proof. Lines through a point are taken by a collineation into lines through the image of the point; further, any line through the image of the point is the image of some line through the original point. A collineation which maps a sphere onto a sphere maps a line intersecting the original sphere onto a line intersecting the image of the sphere; the image line intersects the image sphere at the images of the original points of intersection. Points interior to a sphere, and no other points, have the property that any line through them intersects the sphere at two different points. Hence the validity of the theorem is clear.

Corollaries. *A collineation which maps a sphere onto a sphere maps tangent planes onto tangent planes and tangent lines onto tangent lines; therefore the image of a circular tangent cone to the original sphere is a circular tangent cone to the image sphere.*

Theorem 2. *A collineation which maps a sphere onto a sphere induces a circle- and angle-preserving mapping on the spheres.*

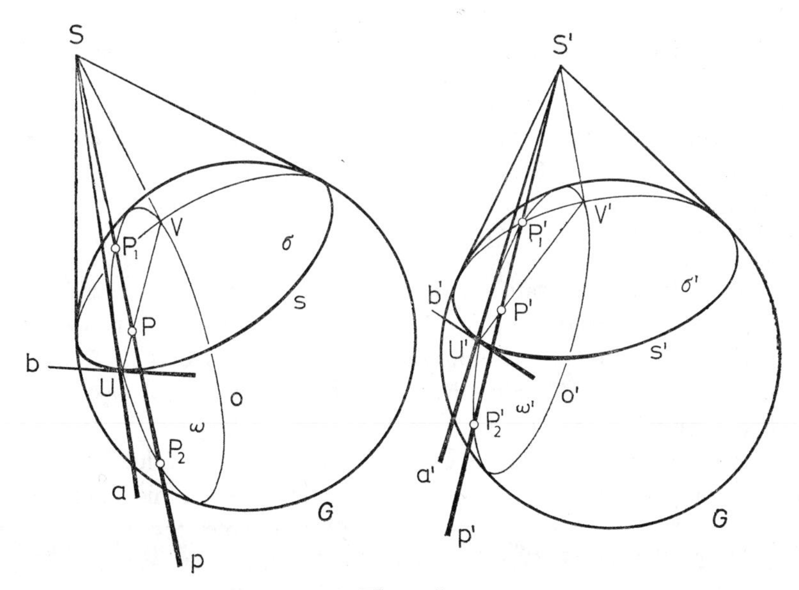

Figure 8

Proof. The mapping induced on the spheres preserves circles since the image of a plane section is a plane section. It remains to prove that a collineation which maps a sphere onto a sphere maps a pair of tangent lines meeting at a point of the sphere onto a pair of tangent lines having the same property relative to the other sphere and forming an angle of equal size.

Let the plane ε which is spanned by the mutually perpendicular lines a, b touch **G** at the common point U of a and b. If the circular cone formed by the tangents drawn to **G** from an arbitrary point S of a has curve of contact s, then b is the tangent of s at point U (Fig. 8).

According to what we have said in connection with Theorem 1, s' touches b' at U', and to the circular cone $S(s)$ there corresponds the circular cone $S'(s')$. Line a' is a generator of the cone $S'(s')$, and b' is a tangent to circle s' of the cone; thus $a' \perp b'$.

Therefore, a collineation **T** which maps a sphere onto a sphere induces a mapping on the spheres that preserves circles and right angles. As a consequence, **T** preserves all angles.

For let a, b, c, d and a', b', c', d' denote four elements of a pencil of lines tangent to **G** and their images, respectively. The latter quadruple of lines is tangent to **G**$'$ and for the cross ratios we have

$$(abcd) = (a'b'c'd').$$

We already know that $a \perp b$ and $c \perp d$ imply $a' \perp b'$ and $c' \perp d'$. Then, however, $(abcd) = -\mathrm{tg}^2\,(ac)$, so that from the equation above

$$\sphericalangle(ac) = \sphericalangle(a'c').$$

We have proved that the mapping induced (on the spheres) by **T** is angle-preserving.

Theorem 3. *Let s and s' be arbitrary circles on the spheres* **G** *and* **G**$'$*, respectively; denote the planes of these circles by σ and σ'. Suppose further that O and O' are points in σ and σ' interior to* **G** *and* **G**$'$*, respectively, while U and U' are points on s and s'. Among all collineations which map* **G** *onto* **G**$'$ *there are four that take the elements s, O and U into s', O' and U', respectively* (Fig. 9).

Proof. If the collineation **T** satisfies the conditions of the theorem, then between the planes σ and σ' it induces a collineation $\mathbf{T}_0$ which takes points interior to s into points interior to s' and takes the elements s, O and U into the elements s', O' and U', respectively. Such collineation **T** really exists; in fact, there are two (plane configurations in σ and σ' are set off on Fig. 10).

For, $\mathbf{T}_0$ satisfying the conditions takes the common point V of line UO and circle s into the common point V' of $U'O'$ and s', and the tangents UH and VH into the tangents $U'H'$ and $V'H'$. Therefore $\mathbf{T}(H) = H'$ ($\mathbf{T}(x)$ denotes the image of figure x under the transformation **T**). Consequently, $\mathbf{T}_0$ takes the common points Y_1 and Y_2 of OH and s into the common points Y_1' an d Y_2' of $O'H'$ and s'. Two cases may

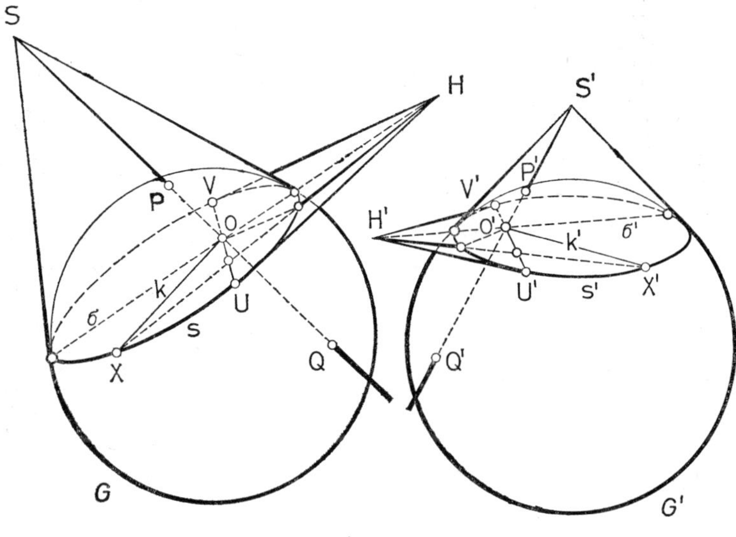

Figure 9

occur:

$$\text{either } \mathbf{T}_0(Y_1)=Y_1' \text{ and then } \mathbf{T}_0(Y_2)=Y_2',$$

$$\text{or } \mathbf{T}_0(Y_1)=Y_2' \text{ and then } \mathbf{T}_0(Y_2)=Y_1'.$$

Both cases are not only possible but also realizable in one and only one way. In fact, it is well known that a collineation of the plane is uniquely determined by the images of four points no three of which are collinear. So, there is exactly one collineation taking U, V, H and Y_1 into, say, U', V', H' and Y_1'. This collineation takes circle s into circle s' since collineations take conics into conics and the position of a conic is determined in a one-to-one manner by three points (H', V' and Y_1') and the tangents ($H'U'$ and $H'V'$) at two of these points. According to the two possible choices for Y_1' and Y_2', the image of the dark segment of the circular domain σ will be either the dark or the light segment of σ'.

One sees easily that the following theorem is true: If $\mathbf{K}_0$ is a collineation between the planes ω and ω', while the points A and B, not in ω, and the points A' and B', not in ω', have the property that the points P and P' in which the line AB meets ω and the line $A'B'$ meets ω', respectively, are associated with each other by the collineation $\mathbf{K}_0$, then there is a unique extension of $\mathbf{K}_0$ to a collineation $\mathbf{K}$ of the whole space which satisfies the relations

$$\mathbf{K}(A) = A', \quad \mathbf{K}(B) = B'.$$

Indeed, it is well known that a collineation of the space is uniquely determined by the images of five points no four of which are coplanar. Therefore, if we choose points C, D, E in the plane ω so that no three of P, C, D, E are collinear and we put $\mathbf{K}_0(C)=$

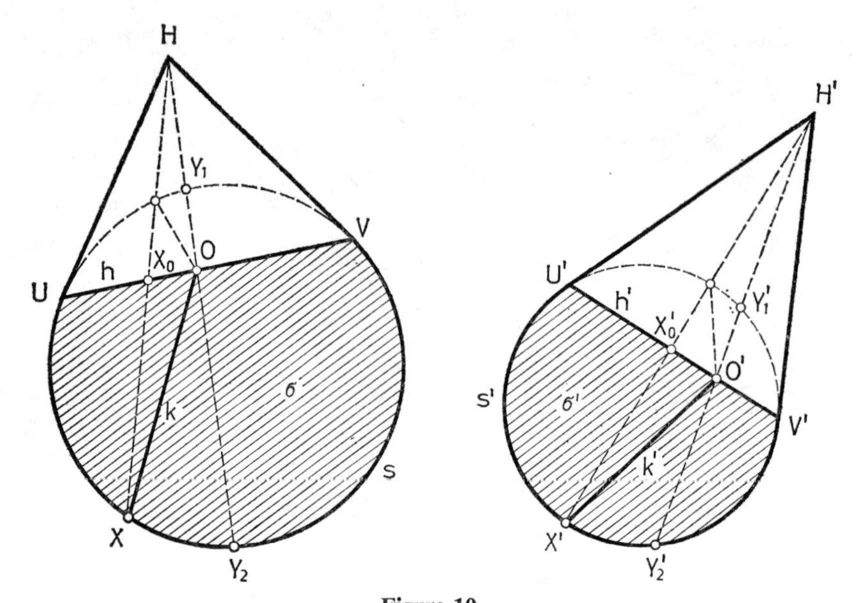

Figure 10

$=C'$, $\mathbf{K}_0(D)=D'$ and $\mathbf{K}_0(E)=E'$, then the points A, B, C, D, E and their images uniquely determine the collineation $\mathbf{K}$ having the required properties.

Now let $\mathbf{T}_0$ and $\mathbf{T}$ appearing above be related to each other as are $\mathbf{K}_0$ and $\mathbf{K}$ just described. If $\mathbf{T}$ has to take $\mathbf{G}$ into $\mathbf{G}'$, then it has to take the pole S of σ with respect to $\mathbf{G}$ into the pole S' of σ' with respect to $\mathbf{G}'$ (according to what we said in connection with Theorem 1). Let the roles of A and A' be played by S and S' (a suitable choice of B and B' will be specified immediately).

If line SO intersects $\mathbf{G}$ at points P and Q, while $S'O'$ intersects $\mathbf{G}'$ at P' and Q', then two cases may occur:

either $\mathbf{T}(P)=P'$ and then $\mathbf{T}(Q)=Q'$,

or $\mathbf{T}(P)=Q'$ and then $\mathbf{T}(Q)=P'$.

So, in the first case P and P', while in the second case P and Q', can play the roles of B and B' (Fig. 9).

Consider, for instance, the first case. The collineation $\mathbf{T}_0$ and the requirements $S\to S'$, $P\to P'$ define a collineation $\mathbf{T}$ of the whole space which maps the sphere $\mathbf{G}$ tangent along s to cone $S(s)$ onto a quadratic surface, and the cone itself onto the circular cone $S'(s')$. This quadratic surface, however, has to be tangent along the circle s' to $S'(s')$. All quadratic surfaces with the latter property form a pencil of surfaces, and to any point, not on s', of the space there is one and only one surface in the pencil passing through it. If this arbitrary point is precisely on $\mathbf{G}'$, then the collineation $\mathbf{T}$ maps $\mathbf{G}$ precisely onto $\mathbf{G}'$.*

* Since the proofs did not require that $\mathbf{G}$ and $\mathbf{G}'$, or s and s', were distinct, Theorems 1–3 remain valid for a single sphere (and possibly a single circle on it).

We now continue the treatment of the model. We show that, in the model space, each axiom of hyperbolic geometry corresponds to a correct (Euclidean) theorem.

The system of axioms for hyperbolic geometry coincides with Hilbert's axioms apart from a discrepancy in IV; so the checking of this axiom will be left to the end of the discussion.

The assertions of axioms I_1-I_8, II_1-II_4 and V_2 remain valid in the portion of space enclosed by a sphere provided that all elements appearing in the axioms are interior to the sphere. Consequently, the *"space"* satisfies these axioms.

By the axioms of congruence, III_1 is equivalent to the following assertion in the model.

Theorem. *Let us be given a sphere* **G**, *points A, B and A' interior to* **G**, *and a chord a' through A'. There is a collineation which maps* **G** *onto itself, A into A', and B into some point of a prescribed section, cut off by A', of a'. Any collineation with these properties maps B into one and the same point* (Fig. 11).

The existence of a collineation having the required properties follows readily from Theorem 3. **G** cust the chord $a=MN$ out of the line AB. Denote the end-points of a' by M' and N'. Through these chords we can take arbitrary planes σ and σ', and we know that there is a collineation **K** for which $\mathbf{K}(\mathbf{G})=\mathbf{G}$, $\mathbf{K}(\sigma)=\sigma'$, $\mathbf{K}(A)=A'$ and, say, $\mathbf{K}(N)=N'$.

Since collineations preserve cross ratios, for the point $\mathbf{K}(B)=B'$ we have

$$(MNAB) = (M'N'A'B').$$

Thus all collineations with the required properties map B into one and the same point B'.

In the model, to III_2 there corresponds the following

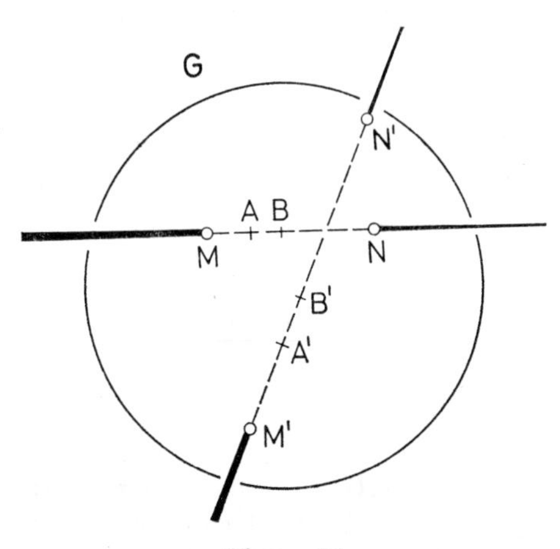

Figure 11

Theorem. *Let the segments AB, $A'B'$, $A''B''$ lie interior to the sphere* **G**. *If among the collineations which map the sphere onto itself there is one that maps $A'B'$ onto AB and one that maps $A''B''$ onto AB, then there is one that maps $A'B'$ onto $A''B''$,*

If $\mathbf{T}_1$ and $\mathbf{T}_2$ are the collineations the existence of which is assumed in the theorem, then from the properties

$$\mathbf{T}_1(\mathbf{G}) = \mathbf{G}, \quad \mathbf{T}_1(A'B') = AB;$$

$$\mathbf{T}_2(\mathbf{G}) = \mathbf{G}, \quad \mathbf{T}_2(A''B'') = AB$$

it follows immediately that

$$\mathbf{T}_2^{-1}(\mathbf{T}_1(\mathbf{G})) = \mathbf{G} \quad \text{and} \quad \mathbf{T}_2^{-1}(\mathbf{T}_1(A'B')) = A''B''.$$

Thus the collineation $\mathbf{T}_2^{-1}\mathbf{T}_1 = \mathbf{T}$ really fits in the conclusion of the theorem.
In the model, to III_3 there corresponds the following

Theorem. *Let us be given the sphere* **G**. *Let ABC and $A'B'C'$ be collinear triples of points interior to the sphere such that B lies between A and C, while B' lies between A' and C'. If among the collineations which map* **G** *onto itself there is one that maps AB onto $A'B'$ and one that maps BC onto $B'C'$, then there is one that maps AC onto $A'C'$.**

The proof is based on the fact that any mapping of a *finite* line segment onto a *finite* line segment which preserves cross ratios preserves also the ordering of the points. Thus the collineation $\mathbf{K}$ for which $\mathbf{K}(AB)=A'B'$ takes C into a point $\mathbf{K}(C)$ which is separated from A' by B'. Therefore, by what we have said about Axiom III_1, only $\mathbf{K}(C)=C'$ is possible.
In the model, equivalent with III_4 is the following

* At this point we note that, in the model, the role of "*distance*" (measure of a segment) is played by

$$\varkappa|\log(MNAB)| = d(AB),$$

where the value of the parameter $\varkappa > 0$ is arbitrary but the same for all quadruples M, N, A, B. From the relations

$$(MNAA)=1,$$

$$(MNAB) \neq 1 \quad \text{if } A \text{ is different from } B,$$

$$(MNBA)=1:(MNAB),$$

$$(MNAB)\cdot(MNBC)=(MNAC)$$

if and only if B lies between A and C on the chord MN valid and well known for the cross ratio one sees immediately that $d(AB)$ fulfils the requirements which define distances associated with segments of one and the same line. It also follows at once that distances associated with congruent segments, and only those, are equal.

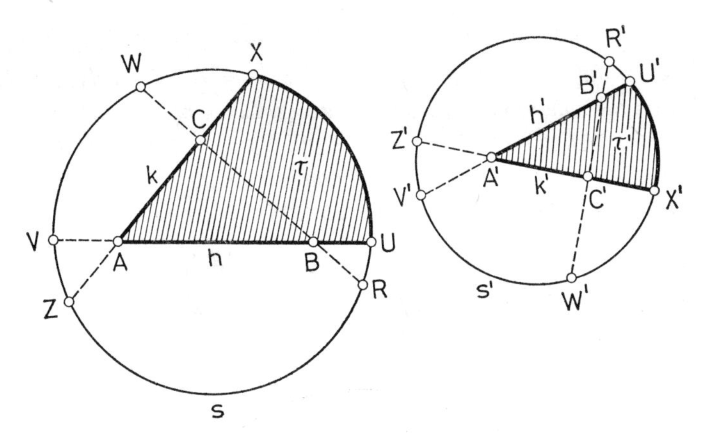

Figure 12

Theorem. *Let U and X be any points of the sphere* **G**, *and O a point interior to the sphere. Let the point V be such that the chord UV contains O. Let O′ be an interior point of the chord that connects the arbitrary points U′ and V′ of* **G** . *The domain enclosed by* **G** *in the plane [OUX] is dissected by OU into two circular segments; denote by σ_1 that segment which contains X. Let s′ be any circle of* **G** *passing through U′ and V′; it is dissected by the chord U′V′ into two segments one of which will be denoted by σ_1' and the corresponding arc by s_1'. There are collineations which map* **G** *onto itself and the points O, U and V into O′, U′ and V′, respectively. Among them there are two which map σ_1 onto σ_1'. Each of these two collineations takes X into one and the same point X′ of the arc s_1'.*

We discussed and established this theorem already when proving Theorem 3 (the roles of σ_1 and σ_1' were played by the dark segments appearing there).

III_5 is equivalent to the following theorem in the model:

Theorem. *Let ABC and A′B′C′ be two triangles interior to the sphere* **G**, *and let their planes intersect* **G** *in s and s′, respectively. If there is a collineation which maps the segment AB onto A′B′, one which maps AC onto A′C′, and one which maps the angular domain CAB onto the angular domain C′A′B′, then there is a collineation which maps the angular domains ABC and BCA onto the angular domains A′B′C′ and B′C′A′, respectively* (Fig. 12).

In fact, according to what was established in connection with III_1, if the collineation **K** satisfies the relation $\mathbf{K}(\sphericalangle CAB) = \sphericalangle C'A'B'$ then $\mathbf{K}(B)=B'$ and $\mathbf{K}(C)=C'$. Thus for the same collineation also

$$\mathbf{K}(\sphericalangle ABC) = \sphericalangle A'B'C' \quad \text{and} \quad \mathbf{K}(\sphericalangle BCA) = \sphericalangle B'C'A'.$$

Finally, we check the reformulation of V_1 in the model. It reads as follows (see Fig. 7):

Theorem. *Let AB and CD be any segments interior to the sphere* **G**. *Let the line AB intersect* **G** *at U on the side of A, and at V on the side of B. Similarly, let CD intersect* **G** *at M and N on the side of C and D, respectively. Let* **T** *be a collineation which maps* **G** *onto itself and satisfies the relations $\mathbf{T}(M, C, N) = U, A, V$ (then the image of D under* **T**, *say A_1, lies between A and V). Finally, let $\mathbf{T}_0$ be a collineation which maps* **G** *onto itself, U and V into themselves, and A into A_1. The points*

$$A_0 = A, \quad A_1 = \mathbf{T}_0(A), \quad A_2 = \mathbf{T}_0(A_1), \ldots$$

proceed from A towards V, and there is a (finite) index k such that B lies between A and A_k.

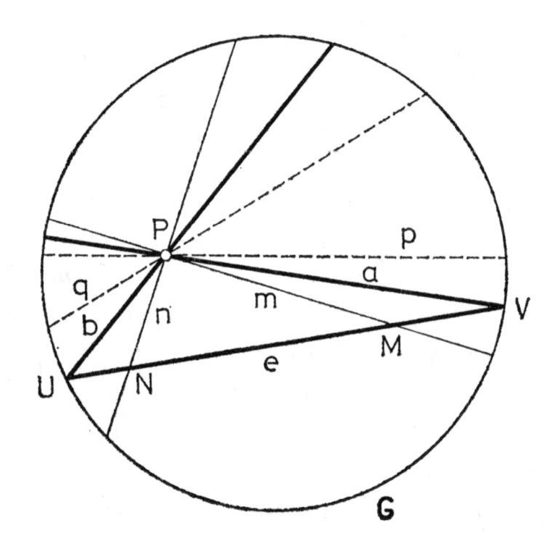

Figure 13

The proof starts from the fact that, owing to the configuration of the points, the cross ratio

$$(MNCD) = \lambda$$

is positive and less than 1. Using our notation,

$$(UVAA_1) = (UVA_1A_2) = \ldots = (UVA_{k-1}A_k) = \lambda.$$

Hence

$$(UVAA_1)(UVA_1A_2)\ldots(UVA_{k-1}A_k) = (UVAA_k) = \lambda^k.$$

On the other hand, the value $(UVAB) = \mu$ is also positive and less than 1. So

$$\lambda^k < \mu$$

if k is sufficiently large. In the latter case, A_k lies closer to V than does B. The proof is complete.

The axioms checked so far are axioms of Euclidean geometry, they form the non-complete axiom system of absolute geometry, the residual system of axioms. Let us add to them the axiom of hyperbolic geometry (which replaces IV):

There is a line e and a point P not on e such that, in the plane spanned by them, several lines incident with P do not intersect e (Fig. 13).

This axiom means in the model that in the plane spanned by a point P interior to **G** and a chord UV there are several lines through P which intersect the line UV outside the sphere. The latter statement is really obvious. Thus we have finished proving that hyperbolic geometry is consistent (provided that Euclidean geometry is so).*

8. POINCARÉ'S MODEL

We describe *Poincaré*'s model of hyperbolic geometry. It was constructed in 1882 by Poincaré who was led to the basic idea of the model by his investigations in function theory. A great advantage of this model (over the **CK**-model** treated above) is that *congruent* angles of hyperbolic space are represented by *equal* angles in the model.

Before describing the model of solid geometry it is necessary to explain that of plane geometry. We derive the **P**-model of plane geometry from the **CK**-model of plane geometry. Consider the **CK**-model of hyperbolic plane geometry in the circle h (Fig. 14). Choose one of the hemispheres, bounded by h, of the sphere determined by the great circle h. First, associate with each point interior to h — and belonging there-

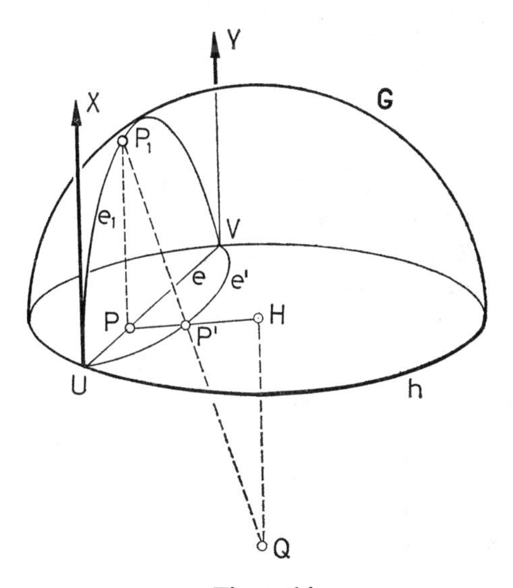

Figure 14

* The model used in the proof was created for representing hyperbolic geometry by F. Klein (1871) who drew the idea from A. Cayley's paper *A sixth memoir upon quantics* (1859).

** For brevity, the Cayley–Klein model and the Poincaré model will be written as **CK**-model and **P**-model, respectively.

fore to the **CK**-model — a point of this hemisphere. Namely, if P is a point in the **CK**-model, then we associate with it that point P_1 of the hemisphere at which the perpendicular erected at P to the plane of h intersects the hemisphere. Let Q denote the point at which the diameter perpendicular to the plane of h intersects the complementary hemisphere. Project the points of the original hemisphere from Q to the plane of h (the projection of the projection P_1 appearing above will be some point P'). The two consecutive projections carry each point of the **CK**-model into a point of the same disc (the correspondence is one-to-one). This correspondence defines a new model of the hyperbolic plane, the **P**-model.

We observe that in the **P**-model *the role of straight lines is played by the arcs interior to h of those circles which intersect h perpendicularly*. For, projecting the chords of the **CK**-model to the hemisphere yields semi-circles which intersect h perpendicularly (for instance, the semi-circle e_1 corresponding to chord e on Fig. 14). Projection from Q to the plane of h is circle- and angle-preserving.* So, in fact, this projection takes those circular arcs of the hemisphere which intersect h perpendicularly into circular arcs interior to h which intersect h perpendicularly in the plane of h.

Now we can define the spatial model. Consider the **CK**-model given by a sphere **G**, and let P be a point interior to **G**. Take a great circle h of **G** the plane of which contains P. In the **P**-model determined by this circle h, find the point P' corresponding to P in the way described above. From the rotational symmetry of the sphere it follows that P' does not depend on the choice of h and, therefore, is uniquely determined by P. The point P' is called the point associated in the spatial **P**-model with point P of the **CK**-model.

Thus the **P**-model defined by one and the same sphere **G** can be derived from the **CK**-model with the help of a one-to-one point transformation. *In the* **P**-*model, the roles of straight lines and planes, respectively, are played by arcs interior to* **G** *of circles that intersect* **G** *perpendicularly and caps interior to* **G** *of spheres that intersect* **G** *perpendicularly*. Indeed, consider a chord of **G** and the great circle h that passes through its end-points. According to what we established in the previous paragraph, the images of the points of this chord can be constructed, for instance, by means of the planar **P**-model determined by the great circle h. We already know that the collection of all points constructed in this way coincides with the interior arc of a circle which lies in the plane of h and intersects h — and therefore **G** — perpendicularly.

On the other hand, the validity of our assertion concerning the counterpart of the plane in the **G**-model follows from the assertion concerning the counterpart of the line just proved. For, the plane of any circle of the sphere **G** can be obtained by rotation of a chord about the diameter that intersects this chord perpendicularly. This rotation is at the same time a rotation of the arc corresponding to the chord in the

* The plane of h is perpendicular to the diameter through Q of the sphere. Therefore, the plane of h and the point Q are the range and centre of a stereographic projection system. The stereographic projection of a sphere, however, is circle- and angle-preserving.

P-model by which this arc describes a spherical cap interior to **G** and intersecting **G** as well as the axis of revolution perpendicularly.

Incidence and betweenness retain their meaning in the **P**-model. On the other hand, the definition of congruence in the **P**-model should be treated in some detail. We want to show that the role of congruent angles in the **P**-model is played by equal angles.

We first deal with angles in the planar **P**-model determined by the circle h. In the corresponding **CK**-model the roles of angle and angular domain are played by a chord configuration UAV interior to h and the part interior to h of the Euclidean angular domain UAV, respectively. Let the angular domain of the **CK**-model be denoted by Λ for short. Projecting it up, perpendicularly to the plane of h, to one of the hemispheres of **G** bounded by h we there obtain a domain Λ_1 which is bounded by circular arcs $\widehat{A_1U}$ and $\widehat{A_1V}$ intersecting the great circle h perpendicularly (Fig. 15*). The tangents at A_1 of the arcs intersect the plane of h at points M and N. If we project the spherical domain Λ_1 back to the stereographic plane h from the corresponding — lying on the complementary hemisphere — stereographic centre Q, then we obtain a domain Λ' which is bounded by circular arcs $\widehat{A'U}$ and $\widehat{A'V}$ intersecting h perpendicularly; this is the angular domain corresponding to Λ in the **P**-model. Since stereographic projection preserves angles, the spherical angle MA_1N and the projected angle $MA'N$ are equal in the Euclidean sense (Fig. 16).

In the **CK**-model determined by the circle h, Λ and Λ^* signify congruent angles if there is a collineation $\mathbf{T}_0$ which maps h onto itself and Λ onto Λ^*. According to what

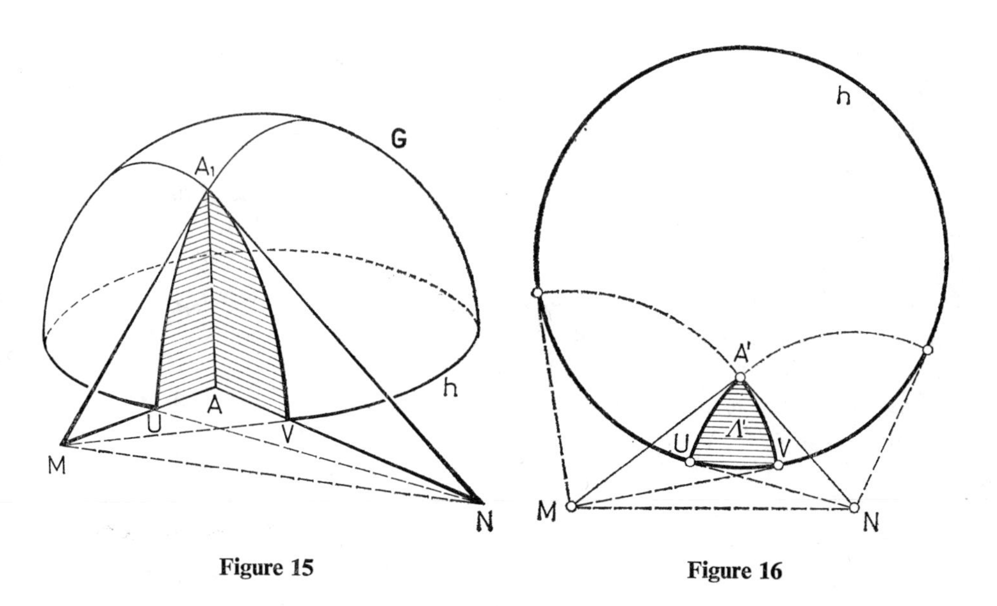

Figure 15

Figure 16

* On the figure, this domain is removed and the planes of the arcs made conspicuous by shading.

14

we said in §6 when proving Theorem 3, this collineation can be extended to a collineation **T** of the whole space which maps the sphere **G** determined by h onto itself. Perpendicularly projecting up the chords and points of the plane of h to the sphere **G** may be regarded as the intersection with **G** of the projection planes and projection lines through the vertex (now: the ideal point in the direction perpendicular to the plane of h) of the circular cone (now: circular cylinder) tangent along h. The collineation **T** takes these planes and lines again into planes and lines (projecting Λ^*) that are perpendicular to the plane of h since the vertex of the circular cone tangent along h is invariant under **T**. Consequently, **T** maps Λ_1 (corresponding to Λ on the hemisphere) onto Λ_1^* (corresponding to Λ^* on the hemisphere). According to §6, Theorem 2, **T** induces an angle-preserving mapping on the sphere. Therefore, the curvilinear sectors Λ_1 and Λ_1^* on **G** associated with angles considered congruent in the **CK**-model have equal angles in the Euclidean sense. The curvilinear sectors Λ' and $\Lambda^{*\prime}$ of the **P**-model obtained by stereographic projection also have equal angles, $\sphericalangle M\Lambda'N=$ $=\sphericalangle M^*\Lambda^{*\prime}N^*$, and this was to be proved. Thus *the* **P**-*model is an angle-preserving image of the hyperbolic plane.*

Also, *the spatial* **P**-*model is an angle-preserving image of hyperbolic space,* which we prove by a similar argument. Consider the **CK**-model and **P**-model determined by the sphere **G**. Let h_1 and h_2 be arbitrary circles on **G**. Denote by $\mathbf{G}_1$ and $\mathbf{G}_2$ those spheres which intersect **G** perpendicularly in h_1 and h_2, respectively. Consider two chords $U\bar{U}$ and $V\bar{V}$ of h_1 as well as two chords $U^*\bar{U}^*$ and $V^*\bar{V}^*$ of h_2 named by their end-points. Assume that the first two chords intersect at the point O interior to h_1 while the other two intersect at O^* interior to h_2. The figures UOV and $U^*O^*V^*$, to be denoted by Λ and Λ^* for short, mean two angles when considered in the **CK**-model. We take Λ and Λ^* to be congruent in the **CK**-model provided there is a collineation which maps **G** onto itself and Λ onto Λ^*. From the action of this **T**, consider only that correspondence $\mathbf{T}_0$ which ascribes h_2 to h_1 and Λ^* to Λ; that is, the correspondence which carries the circle h_1 of sphere $\mathbf{G}_1$ and the figure Λ of the plane of this circle into the circle h_2 of sphere $\mathbf{G}_2$ and the figure Λ^* of the plane of the latter circle. This $\mathbf{T}_0$ can be extended to a collineation $\mathbf{T}^*$ of the whole space which takes $\mathbf{G}_1$ into $\mathbf{G}_2$ (§6, Theorem 3). The vertices of the cones tangent to $\mathbf{G}_1$ along h_1 and to $\mathbf{G}_2$ along h_2 coincide with each other and with the centre of **G**. Project Λ to $\mathbf{G}_1$ and Λ^* to $\mathbf{G}_2$ from the centre of **G**. The collineation $\mathbf{T}^*$ takes the centre of **G** into itself; therefore, it takes the lines and planes of projection attached to $\mathbf{G}_1$ into the respective elements of projection attached to $\mathbf{G}_2$ and, consequently, maps the projection obtained on $\mathbf{G}_1$ onto the projection obtained on $\mathbf{G}_2$*. So, if Λ_1 is the projection of Λ from the centre of **G** to the spherical cap of $\mathbf{G}_1$ interior to **G** and Λ_1^* is the projection of Λ^* from the same point to the cap of $\mathbf{G}_2$ interior to **G**, then $\mathbf{T}^*(\Lambda_1)=\Lambda_1^*$. Now, according to §6, Theorem 2, the angles of the spherical arc figures Λ_1 and Λ_1^* must be equal. On the other hand, Λ_1

* We note that although $\mathbf{T}^*$ takes h_1 and Λ (belonging to h_1) into h_2 and Λ^* (belonging to h_2), in general it maps **G** only onto a quadratic surface containing h.

and Λ_1^* are the angle figures in the **P**-model determined by **G** which correspond to the angle figures Λ and Λ^* of the **CK**-model determined by **G**. Thus congruent angles are represented by equal angles in the **P**-model, indeed.

We next describe the counterparts in the **P**-model of a few configurations of hyperbolic geometry.

First the planar **P**-model will be considered. Fig. 17 shows the three kinds of relative position of lines in the plane. The pair *e*, *g* plays the role of intersecting lines, *e* and *d* represent parallel lines, while *d* and *g* represent non-intersecting lines.

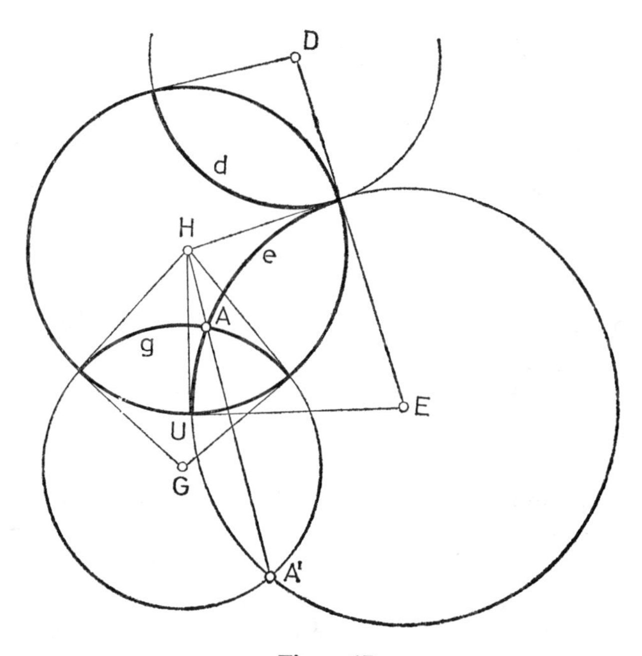

Figure 17

Fig. 18 represents, in the **P**-model, three kinds of a flat pencil of lines realizable in the hyperbolic plane. The first is the flat pencil formed by all lines passing through a point *K*. It is easy to see that, in the **P**-model, the role of this pencil is played by a hyperbolic pencil of circles taken in the Euclidean sense. Really, the null-circle **K** and the circle *h* determine an elliptic pencil of circles, and the circles that intersect the elements of the pencil perpendicularly form the conjugate pencil of circles which is hyperbolic. The pencil consisting of parallel lines is shown on figure b). Obviously, parallel lines form a parabolic pencil of circles in the model. Figure c shows that the collection of all lines perpendicular to a given line is represented by an elliptic pencil of circles. In fact, the circular arc *k* which intersects also *h* at a right angle plays the role of a straight line of the hyperbolic plane. Further, a perpendicular to this line is a circular arc that intersects *k* and *h* perpendicularly. Since, however, *h* and *k* deter-

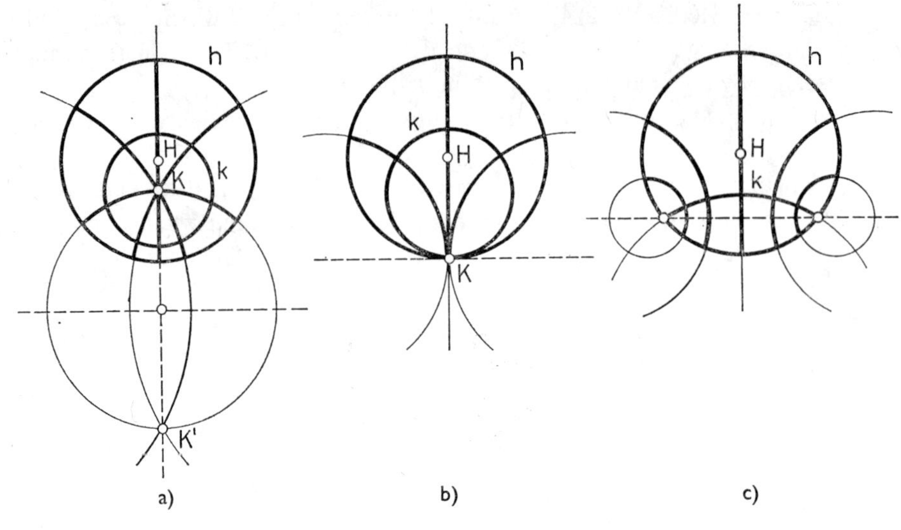

Figure 18

mine a hyperbolic pencil of circles, the perpendicularly intersecting circles form the conjugate, elliptic, pencil of circles.

The cycles of hyperbolic geometry can also be defined by requiring that *the cycle is an orthogonal trajectory of a flat pencil of lines*. Hence it follows immediately what we should regard as a cycle in the **P**-model: an orthogonal trajectory of the pencil of circles representing a pencil of lines, which is a circle again (of course, only the part interior to circle h should be considered). Corresponding to the three kinds of flat pencils of lines we obtain three kinds of cycles in the **P**-model: a circle k completely interior to h (figure a)), a circle k internally tangent to h (figure b)), and the arc interior to h of a circle k that intersects h at two points (figure c)). These circles k correspond to a circle, a paracycle and a hypercycle, respectively.

Recalling that the spatial **P**-model arises from the planar **P**-model by rotation, we realize at once that the roles of sphere, parasphere and hypersphere are taken by spheres in the **P**-model. A sphere is a sphere interior to the sphere **G**; a parasphere is a sphere internally tangent to **G**; a hypersphere is the cap interior to **G** of a sphere that intersects **G** in a circle.

From these considerations it follows that, *in the spatial* **P**-*model, congruence is replaced by a point transformation which carries* **G** *into itself and is sphere- and angle-preserving, while in the planar* **P**-*model it is replaced by a point transformation which carries h into itself and is circle- and angle-preserving*. We prove that every point transformation of this kind corresponds to a congruence.

To this end we only have to show that point transformations with the properties above carry circular arcs which intersect sphere **G** or circle h perpendicularly into circular arcs of the same kind. Really, then the counterpart in the **CK**-model of such a point transformation is a line-preserving point transformation, that is a collineation.

This, in turn, for the planar model follows immediately from the requirements, whereas for the spatial model it is a consequence of the fact that a sphere-preserving mapping takes any intersection of spheres, that is any circle, into an intersection of spheres, which is again a circle.

In the sequel we restrict our attention to the planar **P**-model. A good means for the study of circle- and angle-preserving point transformations which map a circle h onto itself is provided by the theory of functions of a complex variable. Consider the unit circle of the complex plane in the role of h. A mapping between points of the complex plane can be written in the form

$$f(z) = w,$$

where z means the original point, w is the image, and the function is the mapping. The function $f(z)=w$ induces an angle-preserving mapping of the interior of the unit circle if it is *regular* within the unit circle and $f'(z)\neq 0$. Moreover, this mapping is circle-preserving if

$$w = \frac{\alpha z+\beta}{\gamma z+\delta} \quad \text{(where } \alpha\delta-\beta\gamma \neq 0\text{)},$$

that is, if w is a so-called linear fractional function of z. This function assumes an even more special form if we also require that it carry the unit circle into itself and interior points into interior points. In this case,

$$w = \frac{\alpha z+\beta}{\bar{\beta} z+\bar{\alpha}} \quad \text{(where } \alpha\bar{\alpha}-\beta\bar{\beta} > 0\text{)}.$$

So the study of point transformations of the hyperbolic plane which give rise to congruent images reduces to the study of these special linear fractional functions.

The transformations which have a matrix of this special kind form a group (the product of two elements of the group is the transformation obtained by consecutively performing the two transformations serving as factors). This nice and simple example illustrates FELIX KLEIN's famous assertion that the subject of any geometry is the study of geometric properties which are invariant under the transformations belonging to some transformation group. This is the basic content of his celebrated Erlanger Programm published in 1872.

We mention that the idea of relating hyperbolic geometry to the theory of functions of a complex variable played an important role also in the advance of function theory.

CHAPTER III

THE EFFECT OF THE DISCOVERY OF NON-EUCLIDEAN GEOMETRY ON RECENT EVOLUTION OF MATHEMATICS

9. THE FORMATION AND DEVELOPMENT OF THE CONCEPT OF MATHEMATICAL SPACE

Our remarks to the *Appendix* as well as our discussions about the progress of non-Euclidean geometry and about the problem of consistency reflect, first of all, the direct influence that BOLYAI's and LOBACHEVSKY's discovery, their way of mathematical thinking exerted on the development of mathematics and, especially, on the ideas of 19th century mathematicians. This influence, however, has spread further and, although less directly, can be traced also in recent advance of mathematics.

In the 19th century, views on mathematics underwent a radical change. Prior to BOLYAI and LOBACHEVSKY, the concept of mathematical space had not been cleared up at all. Even a sufficiently sharp distinction between the concepts of mathematical and physical space had been missing. It had not been realized yet that geometry is not necessarily the only possible reflection of objective spatial relations existent in the material world, but a tool for further development of the space concept, a tool which besides containing the description of physical space can also reflect other objective relations in the material world. For the formation of this general concept of mathematical space, relations between axioms of Euclidean space had to be clarified first. One had to crush the scholastic rigidity which had arisen from looking at the Euclidean system of axioms as a necessity of divine origin. It had to be shown that Euclidean geometry is a human creation and that the question whether it is applicable to the material world can be decided only by scientific experience. It had to be realized that the system of geometric axioms is capable of evolution and each step of this evolution leads to a deeper understanding of the material world.

The evolution of the mathematical concept of space began with BOLYAI and LOBACHEVSKY. By overcoming the preconception that Euclidean geometry is only proper, they opened the gate to the recognition of abstract space and started a new, rich period in the evolution of mathematical reasoning.

The next decisive step towards the creation of the mathematical concept of space was made by RIEMANN in 1854. He has also contributed to relaxing the restrictions imposed on the space concept by Euclidean geometry. He went further in the non-Euclidean direction and 19th century mathematics owed him its characteristic feature that geometric methods entered the most different branches of mathematics, branches that did not belong to geometry in the original sense. In the middle, and even more in the second half, of the 19th century, geometries of higher dimensions initiated by pioneering investigations of CAYLEY, CAUCHY and GRASSMANN, the study of configurations of constant curvature in spaces of higher dimension also prepared the creation of the most general concept of mathematical space.

One of the crucial steps in this direction consisted in discovering the scientific importance of topological properties of geometrical configurations. In this respect rather considerable is the fact that JÁNOS BOLYAI in the *Raumlehre* and the notes attached to it, made deep assertions about the topological concept of surface. However, interesting new results and problems involving topology could attain the proper level only after a certain maturity of point set theory. As a result of the advance made by the latter, combinatorial and point-set-theoretical investigations arrived at a synthesis. Thus, at the end of the 19th century, the process having started with the discovery of non-Euclidean geometry reached a stage at which the most general concept of mathematical space could be introduced.

At the beginning of our century, F. RIESZ and M. FRÉCHET defined general mathematical space in two different ways.*

Fréchet defined convergence space by axioms on the convergence of point sequences. Further, he introduced the concept of metric space in the well-known way: for any pair P, Q of points of the space, let us be given a number $\overline{PQ}=\overline{QP}\geqq 0$ so that $\overline{PQ}=0$ if and only if P coincides with Q, and that for any triple P, Q, R of points the triangle inequality $\overline{PQ}+\overline{QR}\geqq\overline{PR}$ is satisfied. From his famous thesis for the doctorate, it turned out already that certain relations which had seemed characteristic for Euclidean geometry hold true in point sets of much more general type.

The method of F. RIESZ is more general than that of FRÉCHET, since RIESZ axiomatized directly the notion of accumulation point and thereby arrived at the concept of the most general topological space. In the course of subsequent development, the method of RIESZ has become the foundation of modern topology. According to RIESZ, *any set X may be considered a space if the notion of accumulation point is defined in it so that the following four axioms are satisfied:*

1. *If P is an accumulation point of the set A, then it is an accumulation point of any set containing A.*

2. *If P is an accumulation point of the set $A\cup B$, then it is an accumulation point of either A or B.*

3. *One-point sets have no accumulation point.*

4. *If P and Q are two distinct accumulation points of the set A, then A has a subset B such that P is, but Q is not, an accumulation point of B.*

* F. RIESZ: *Genesis of the Space Concept* (in Hungarian; Math. Phys. Lapok, Vol. 15, 1906 and Vol. 16, 1907). — *Stetigkeit und abstrakte Mengenlehre* (Atti del IV. Congresso Internazoinale dei Matematici, Roma, 1908, Vol. 2).

M. FRÉCHET: *Sur quelques points du calcul fonctionnel* (Rendiconti Circ. Mat. Palermo, Vol. 22, 1906).

As a result of the activity of FRÉCHET and RIESZ, the most general concept of abstract space has evolved: the space is a set the elements of which are regarded as points, and the organizing principle that rules the set divides these points into two groups, those of accumulation points and isolated points. The topology of the abstract space is determined by the axioms that characterize the organizing principle providing the notion of accumulation point.

The notion of accumulation point, as introduced by RIESZ, has served as a starting point for diverse axiomatic variants of topological space thoroughly discussed in works of HAUSDORFF, SIERPINSKI, KURATOWSKI, MOORE, VIETORIS, TIETZE, ALEXANDROFF, URYSOHN and several other authors.

Owing to the abstract mathematical concept of space, intensive study of function spaces has become possible, and this led to the birth of Hilbert space and functional analysis. According to a celebrated theorem of Urysohn, there is an intimate connection between abstract topological space and Hilbert's coordinate space. To state Urysohn's theorem, the following definitions are needed.

The topological space X is said to have a *countable basis* if a countable number of sets (this will be the countable basis) can be chosen in it so that, regarding these sets as open neighbourhoods, the topology determined by them coincides with the original topology of X. The space X is said to be *normal* if for any closed subsets A and B with no common point there are two open sets G_A and G_B with no common point such that $A \subset G_A$ and $B \subset G_B$. The set of all points in Hilbert space the coordinates of which satisfy the relations $0 \leqq x_n \leqq \frac{1}{n}$ $(n=1, 2, 3, \ldots)$ is called the *Hilbert cube.*

Urysohn's theorem:*

Every normal space with countable basis can be mapped topologically onto a subset of the Hilbert cube.

So, subsets of Hilbert space are topologically equivalent to all spaces in which the topology is consistent with the simplest intuitive requirements.

Urysohn's theorem has opened the way for a method, synthetic in some sense, of constructing non-Euclidean geometry. In fact, the theorem implies that all spaces suitable for geometric investigations can be looked upon as if they were subsets of Hilbert's coordinate space. Let us consider those of the subsets just mentioned which are convex in the following sense of the word:

1. *To any pair of points P, Q there is one and only one point R such that* $\overline{PR} = \overline{QR} = \frac{1}{2}\overline{PQ}$.

2. *To any pair of points P, Q there is one and only one point R such that* $\overline{PQ} = \overline{QR} = \frac{1}{2}\overline{PR}$.

* P. URYSOHN: *Der Hilbertsche Raum als Urbild der metrischen Räume* (Math. Ann., Vol. 92, 1924).

A significant property of sets convex in this sense is that any two points of them can be connected by one and only one continuum which is *isometric* to a line.* (If a metric space can be mapped topologically onto another metric space so that the distance of any two points is equal to the distance of their images, each of the distances being understood in the metric of the respective space, then the two metric spaces are said to be isometric to each other.)

It may be expected that in sets which are convex in the sense described above, some kind of non-Euclidean geometry is valid. An example of such a non-Euclidean geometry is the geometry with Minkowski metric.** BUSEMANN has proved that *the class of all 3-dimensional non-Euclidean spaces with Minkowski metric coincides with the class of those spaces convex in the above sense and having complete metric in which the distance fulfils one more condition, the so-called limit circle axiom.****

(A metric space is said to have complete metric if every Cauchy sequence of points in it has one and only one accumulation point; here the infinite sequence of points $P_1, P_2, P_3, \dots$ is called a Cauchy sequence if, given any number $\varepsilon > 0$, for sufficiently large n the relations $\overline{P_n P_{n+k}} < \varepsilon$ are satisfied.

Busemann's limit circle axiom reads as follows: if for the sequence of points $P_1, P_2, P_3, \dots$ and the pair of points QR

$$\lim_{n\to\infty} \overline{P_n Q} = \infty, \quad \lim_{n\to\infty} (\overline{P_n Q} - \overline{P_n R}) = 0,$$

then

$$\lim_{n\to\infty} (\overline{P_n Q} - \overline{P_n R_1}) = 0, \quad \lim_{n\to\infty} (\overline{P_n Q} - \overline{P_n R_2}) = 0,$$

$$\lim_{n\to\infty} (\overline{P_n Q} - \overline{P_n R_3}) = 0,$$

where R_1 denotes the mid-point of QR, while R_2 and R_3 are defined by R and Q being the mid-points of QR_2 and RR_3, respectively.)

In this brief survey, we have only tried to demonstrate the tremendous progress brought about by the general concept of space. First, with the aid of an organizing principle, abstract sets have developed into spaces suitable for geometric investigations. Second, geometry has become constructible from the elementary notions of set and distance, in contradistinction to the classic method which had deduced geometry from properties — taken for granted *a priori* — of points, lines, and planes. *The great merit of János Bolyai's discovery is that it has opened this immense perspective and made possible the corresponding advance of science.*

* K. MENGER: *Untersuchungen über allgemeine Metrik* (Math. Ann., Vol. 100, 1928).
** H. MINKOWSKI: *Geometrie der Zahlen* (Leipzig, 1910).
*** H. BUSEMANN: *Über die Geometrien, in denen die "Kreise mit unendlichem Radius" die kürzesten Linien sind* (Math. Ann., Vol. 106, 1932).

10. AXIOMATIC METHOD AND MODERN MATHEMATICS

Former appreciations of BOLYAI's and LOBACHEVSKY's discovery, of their science-renewing views, stressed the victory of pure logic over constructive reasoning directed by intuition. The study of the evolution of mathematics in the last one and a half century, however, leads to different patterns of valuation. It is indisputable that, at the beginning of the 19th century, mathematics reached a stage where its fields of problems — because of their diverse and intricate interrelations — could no longer be ruled, and the reliability of the results decided, without consistent use of the axiomatic method. Also unquestionably, the first attempts at axiomatization have been made in geometry. It was the axiomatic treatment which revealed that besides Euclid's there exist a lot of geometries, from various points of view the one more interesting than the other. This idea and the creation of the first non-Euclidean geometry have been the principal merits of both pioneers.

HILBERT's activity, starting from geometry, directly reflects the influence of BOLYAI and LOBACHEVSKY. Proof theory is a fruit of his investigations into the independence of axioms and consistency of the axiom system. A famous result of proof theory is GÖDEL's theorem* from the year 1931:

In a fixed system of axioms it is always possible to find an assertion the correctness or falseness of which cannot be decided.

(Of course, this does not exclude the possibility that the assertion is decidable in another system of axioms.) Since the substance of intuitively learned geometrical concepts cannot be grasped by any number of axioms, the axiomatic method supplements and stabilizes rather than replaces the mathematician's constructive intuition. One should agree with KAGAN that in the work of the discoverers of non-Euclidean geometry it is just the strength of constructive intuition which fascinates the reader.

The modern form of the axiomatic method — undoubtedly, under the influence of BOLYAI's and LOBACHEVSKY's work — has taken shape in the publications of HILBERT. The method gained the approval of contemporary mathematicians within a short time, and this led first to the renewal of algebra and then to the rebuilding of all mathematics. In the proof of a theorem, only a few features of the objects of mathematical research play a role; so the proof can be applied to other objects also having these features. Using this idea, the proof of an assertion is performed without specifying the objects involved; instead, those properties of the objects on which the proof is based are listed as axioms. A theorem proved in this way will be valid for all objects which satisfy the axioms.

Working in this spirit, several fields of mathematical problems considered previously to be separate and remote from one another could be treated alike as particular cases of a more comprehensive problem. On the other hand, constructive intuition

* K. GÖDEL: *Über formal unentscheidbare Sätze der Principia Mathematica und verwandter Systeme*, I (Monatshefte Math. Phys., Vol. 38, 1931).

led to the discovery of rather interesting special cases, and of new problems outside even the enlarged field of questions.

As the axiomatic method spread to all territories of mathematics, classic limits of subdivision dimmed. New classification and reordering of the basic mathematical notions were directed by the ideas of structure theory. Mathematics has thus become more uniform. The deepening of abstraction, the widening of generalization promoted science further and further in exploring reality.

Although this last chapter provides only a sketch of recent progress in mathematics, it may have convinced the reader that JÁNOS BOLYAI had written with rightful pride:

"... I have created a new, different world from nothing".

SUPPLEMENT*

BY BARNA SZÉNÁSSY

All information which might interest the reader could not be built in the previous chapters of this book without breaking the structural unity and violating the train of ideas of the *Appendix*. I have therefore gladly agreed to write a supplement that would serve mainly as a summary of the historical facts. I have considered it my task to report on the most important biographical data and give the details of our, parly recently acquired, knowledge connected with the birth and subsequent history of the Appendix. The two Bolyai's have left behind a rather considerable amount of manuscripts, the greater part of which is to be found at the *Teleki–Bolyai Library* (Marosvásárhely, now Tîrgu-Mureş, Roumania); ample material is kept also in the *Collection of Manuscripts* at the *Hungarian Academy of Sciences* (Budapest). This bequest is still under deciphering; highly valuable writings have been recently published by S. BENKŐ [12]. In possession of them, many things can now be seen clearer and in some respects we must even modify the knowledge acquired previously in connection with the two Bolyai's. As to the structure of this Supplement, I note that I have tried — as far as possible — to avoid repetition of the main text. Literature on the history of the formation of space theories would nowadays make up a whole library. There are also hundreds of works dealing with the life and scientific activity of János Bolyai. At the end of the Supplement I mention only a few essays out of the abundant literature.

*

For understanding the work of JÁNOS BOLYAI it is necessary to refer in some words to the life and mathematical activity of his father FARKAS BOLYAI.

FARKAS BOLYAI was born on 9th February 1775 in the small village Bolya in Transylvania (now: Buia, Roumania) in the family of a smallholder. After receiving elementary and secondary education, the extraordinarily talented young man — as it was usual at that time — widened his knowledge at foreign universities. In 1796 he

* Numbers in ( ) or [ ] refer to pages of this book or to works listed at the end of the Supplement, respectively.

spent a semester in Jena. From September 1796 till June 1799 he pursued his studies in Göttingen. In Göttingen a rather close friendship attached him to GAUSS (1777–1855), also active there. When being together, they often discussed the current problems of mathematics and philosophy. With the departure of GAUSS from Göttingen there began an exchange of letters which lasted more than a half century (1797–1853) in the first years of which often, then at longer and longer intervals, they informed each other about the course of their lives and, occasionally, about mathematical problems. This correspondence (excepting four letters published later) appeared in print [3] in 1899. The whole material contains 24 letters written by FARKAS BOLYAI and 22 letters by GAUSS.

Soon after returning from Göttingen, FARKAS BOLYAI got married. As a fruit of this marriage, JÁNOS BOLYAI was born in Kolozsvár (now: Cluj-Napoca, Roumania) on 15th December 1802.

In 1804, FARKAS BOLYAI became professor of mathematics, physics and chemistry at the Reformed College of Marosvásárhely. There he lived and worked until his death on 20th November 1856.

From his early youth, FARKAS BOLYAI was much concerned with questions related to the Euclidean axiom of parallelism. In this connection he pursued very sharp-witted investigations which, however, have not essentially solved the problem. In two hand-written essays (1804 and 1808; published in [5], Vol. II, pp. 5–22) he attempted to verify, in an indirect way, Clavio's substitute axiom according to which the distance line of the straight line is a straight line. He sent the essays for criticism to GAUSS who soon made a comment on the first one pointing out the defectiveness of the proof ([3], pp. 82–83). In spite of this failure, FARKAS BOLYAI pursued his investigations into the subject. However, he saw more and more convincingly that the solution of the problem could not be obtained by his procedure. Making digressions to various branches of science, and also to fiction, he returned to the problem again and again, but after a certain time he was mostly looking for postulates that would serve as a more intuitive substitute for the Euclidean axiom of parallelism. From the equivalent axioms he has found, best known in the world literature is the following: three points lie on either a line or a circle.

Far from the scientific world, FARKAS BOLYAI had only one possibility to publish his mathematical results: a series of textbooks written for the college. The most significant of them is the thick two-volume work in Latin, usually referred to as the *"Tentamen"* after the introductory word of its title (first edition 1832 and 1833, second edition 1896 and 1904).

In the books of FARKAS BOLYAI one finds several original results which exceed the views of that time. These ideas are related to different branches of mathematics: mathematical logic, set theory, analysis, algebra, the foundations of geometry, etc. The world literature often refers to his root approximation method. Also important are some convergence criteria for infinite series of positive terms due to him. Well known are his definition of "end-like equality of areas" (two plane areas are end-like equal if they can be divided into a finite number of pairwise congruent pieces) and several theorems related to it. Afterwards, numerous mathematicians have dealt with extending, generalizing this notion. Also DAVID HILBERT treats it at length in his work *Die Grundlagen der Geometrie*.

It is a widely accepted opinion that FARKAS BOLYAI was the first mathematician in Hungary to have original results.

*

JÁNOS BOLYAI received elementary and secondary education at Marosvásárhely. His mathematical talent appeared very early. After the final examination, from 1818 he continued his studies in Vienna at the Academy of Military Engineering. He there greatly distinguished himself from his classmates by his mathematical knowledge, but the Academy could not stimulate or assist him in original research work. Nevertheless, it was already in Vienna that he began his meditations on the Euclidean axiom of parallelism — presumably under the influence of his father's activity.

In September 1823, after finishing his studies in military engineering with excellent marks, he was placed to Temesvár (now: Timişoara, Roumania) as a second-lieutenant. From there, on 3rd November 1823, he wrote his father his frequently quoted letter according to one line of which he had created "...a new, different world from nothing". From the contents of this letter it turns out that the main ideas of his geometrical system were already ripe by that time, but we cannot answer the question how far he had got in the details of research till then.

*

Breaking the enumeration of biographical data, we now consider the question of how his geometrical system arose in something more detail, making use also of recently discovered facts.

As noted before, JÁNOS BOLYAI thought of laying a more rigorous foundation of geometry as early as at the age of 18. First, however, he followed the example of his father and tried to prove the axiom of parallelism. Despite initial failure and anxious letters from his father he continued speculation and soon — to our knowledge as early as in 1820 — the idea began to ripen in him that Euclidean geometry is not the only possible system. As a verification we add a picture to be found in his exercise-book on mechanics from the academic year 1820/21. According to what was mentioned at another place of the present volume, these figures are very informative for the specialist since they are the first improvised illustrations of some thoughts that were later developed in the Appendix. A further information on which the search can be based is supplied by the sentence in one of his bequeathed manuscripts saying that the crux of the solution cleared up to him at a winter night of 1823 and then he performed the calculations by candlelight through many nights. Also from a few sentences written by him in the evening of his life we may conclude that towards the

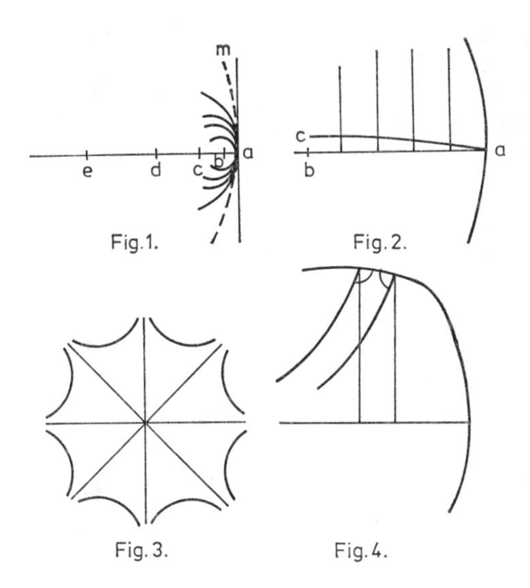

Fig. 1. Fig. 2.

Fig. 3. Fig. 4.

end of 1823 he arrived at the fundamental relation of his geometrical system appearing in §29 of the Appendix. As it seems likely from one of his letters to his father ([11], p. 414), in 1824 the whole material of his treatise might be ready. In fact, from the contents of this letter it follows that at that time he already knew the expression, valid in hyperbolic geometry, for the area of the triangle. This expression, however, is contained in the *last* chapter of the Appendix.

*

The course of JÁNOS BOLYAI's life can be followed exactly. It was usual in the Austro–Hungarian Monarchy of that time to order the officers frequently from one garrison to the other. Thus he also served for a longer or shorter time at Temesvár, Arad, Nagyvárad (now: Oradea, Roumania), Szeged, Lemberg (also Lwów, Ilyvó, now Lvov, USSR) and Olmütz (also Alamóc, now Olomouc, Czechoslovakia). Already from 1826, his health was not perfect; contemporary medical reports refer to malaria, cholera, rheumatic diseases. Recent and rather thorough investigations (Károly Berde) refute the fairly wide-spread assertion as to which he would have suffered from some kind of venereal disease.

Military service could not occupy JÁNOS BOLYAI. He wrote at several places that he was bored by military life as his tasks could have been accomplished by any of his companions: he had to plan barracks, fortresses and similar objects. Tasks of this kind do not help mathematical investigations.

His geometrical results have soon settled to such an extent that in 1825 he could already summarize them. Moreover, he handed over his hand-written treatise (*Raumlehre*; in German) for criticism to his commander in Arad (who had been his military superior at the Vienna Academy of Military Engineering) JOHANN WOLTER von Eckwehr (1791–1857), probably at the beginning of 1826. Unfortunately, we do not know anything about the further fate of that treatise. It might have been destroyed, but the possibility that it lies hidden in some archive cannot be precluded either.

In 1829 the printing of Farkas Bolyai's two-volume work *Tentamen*, in Latin, was authorized. As noted above, it appeared in 1832 and 1833. Urged by the father, János Bolyai drew up again the results of his investigations — in Latin, this time — and handed over the manuscript for publication to Farkas. The short treatise was set up soon and published as an *appendix* at the end of the first volume of the *Tentamen*. A few separata of it, however, were printed previous to the year 1832. They came out of the press on 20th June 1831. One copy was immediately sent to Gauss, but for some reason this consignment has been lost by the post. On 16th January 1832 it was therefore posted again, enclosed with a letter of Farkas Bolyai. In this letter we read the following lines: "Excuse my troubling you — my son appreciates your judgement more than that of all Europe — and is waiting for it all the time. I ask you eagerly to let me know about your opinion"([3], p. 107). This book contains the facsimile of the utmost poorly printed treatise, and a possibly faithful English translation of the Latin text. The internationally accepted title of the treatise, *Appendix,* may strike the reader. The origin of this, actually *wrong,* name is that the two Bolyai's in their correspondence with each other often used, for brevity, the word "Appendix" to specify the treatise, and this led to the spreading of this title later on. When comparing the Latin with the English text the attentive reader may observe a small discrepancy — which does not at all affect the substance of the work: János Bolyai denoted the parameter appearing in his system by i, but later on — for avoiding confusion with the imaginary unit — use of the letter k became generally accepted.

Because of its extreme conciseness — and in spite of the skilfully chosen mathematical symbols — the *Appendix* is difficult to follow. This has made necessary a more detailed treatment of its contents, a reformulation of the proofs in a way easier to understand, and the adoption of a more up-do-date *terminology* which has become current in the course of time.

There are some who explain the laconic style of János Bolyai by financial reasons: namely, the expenses of printing have fallen on him. We know that the two Bolyai's have struggled against financial difficulties throughout their lives; still, economy was not the main reason of the Appendix being concise. In each of his mathematical writings, János Bolyai has *consciously* striven to compose briefly and keep to the point. This is mentioned in his bequeathed manuscripts several times. At one place he writes the following: "It is not the heart of the matter that causes trouble to me, it is rather ... the way how to tell it ... for having the best, the most salutary effect" ([12], p. 35). Unfortunately, this succinct style has meant a handicap for the understanding of the work of János Bolyai as compared to Lobachevsky who composed somewhat easier and developed the proofs in more detail.

*

Having received the *Appendix,* GAUSS replied without delay. He sent to FARKAS BOLYAI a letter which was often quoted later on; its most important parts can be found elsewhere in this book (p. 34).

Objectivity prompts us to report that somewhat earlier, on 14th February 1832, GAUSS had written a few lines of similar content to his former student and friend C. L. GERLING (1788–1864), university professor in Marburg at that time. As this letter is less known, we make the reader acquainted with the original German text of the few lines that are related to the Appendix ([8], p. 387): "Noch bemerke ich, dass ich dieser Tage eine kleine Schrift aus Ungarn über die Nicht-Euklidische Geometrie erhalten habe worin ich alle *meine eigenen Ideen und Resultate* wiederfinde, mit grosser Eleganz entwickelt, obwohl in einer für jemand, dem die Sache fremd ist, wegen der Konzentrierung etwas schwer zu folgenden Form. Der Verfasser ist ein *sehr* junger österreichischer Offizier, Sohn eines Jugendfreundes von mir, mit dem ich 1798 mich oft über die Sache unterhalten hatte, wiewohl damals meine Ideen noch viel weiter von der Ausbildung und Reife entfernt waren, die sie durch das eigne Nachdenken dieses jungen Mannes erhalten haben. Ich halte diesen jungen Geometer v. Bolyai für ein Genie erster Grösse."

The sentences to be read in these two letters of GAUSS were not intended for publicity and, apart from them, the name of JÁNOS BOLYAI does not appear in the bequest of GAUSS. This reserved attitude of GAUSS is very regrettable from the view-point of the history of science, all the more since he often gave account of much less significant works in the columns of the journal "Göttingen Gelehrte Anzeigen".

The manuscripts left behind show that the letter of GAUSS filled the father with pride, and the son with suspicion and indignation. JÁNOS BOLYAI felt that the lines written by GAUSS queried the priority of his discovery. This suspicion upset him very much and he sought another way to have his mathematical results judged. On 8th August 1832, therefore, he wrote a request to his former patron, commander-in-chief of the Vienna Academy of Military Engineering, archduke JOHANN VON HABSBURG. He asked to have his treatise refered by somebody and to get furlough for further research work: he intended to complement and develop the material of the *Appendix.* The request for a furlough was rejected though he needed a rest also because of his steadily deteriorating state of health. Undoubtedly, two referee's reports formed the basis of the unfavourable decision. The report of G. A. GREISINGER, professor of the Academy of Military Engineering, shows that he has not at all understood the *Appendix;* in addition, he lessened the value even of the few words of appreciation to be found in the letter of GAUSS inasmuch as he ascribed them to the early friendship between FARKAS BOLYAI and GAUSS. JOHANN VON HABSBURG was not satisfied with GREISINGER's answer, so he soon asked a new referee: professor A. ETTINGSHAUSEN (1796–1878) from Vienna. Unfortunately, we do not know ETTINGSHAUSEN's report — it may also be discovered somewhere —, but that it was not appreciative can be taken from the following sentence appearing in a letter of JÁNOS BOLYAI from the year 1855: "I esteem Ettingshausen as an excellent and distinguished gentleman, although he is so unlucky, blinded and prejudiced that he cannot appreciate us" ([5], Vol. I, p. 229).

Curiously, the Hungarian Academy of Sciences elected GAUSS (mainly for his results in astronomy and geodesy) in 1847 and ETTINGSHAUSEN in 1858 to a foreign membership, whereas we could not even by careful examination — find the name of JÁNOS BOLYAI in miscellaneous publications of the Academy prior to 1868. Thus, life has denied him due recognition. To be more exact, his share was only what he received from his father and, also, what was contained in the two private letters of GAUSS.

*

The contents of the response letter of GAUSS allow some conclusions. We make a brief digression on them.

1. It is unquestionable that he has carefully examined the *Appendix*, understood its contents, and seen the importance of that work.

2. He has meditated on the philosophical implications of non-Euclidean geometry, in particular about that it cannot be decided *a priori* whether the real world is described by Euclidean geometry or some other one. In other words: the question about the geometric structure of the real world can only be answered through practice. By the way, the two BOLYAI's as well as LOBACHEVSKY have advocated that same principle. We present only one quotation from the manuscripts left behind by JÁNOS BOLYAI ([11], p. 353) in order to demonstrate that he has seen clearly the connection between geometry and physics: "... also the law of gravitation seems to be intimately related... to the nature, essence, construction, quality of space...".

From the mosaic of the hand-written notes left behind by GAUSS and the letters published after his death, researchers could draw up exactly what results he has achieved in hyperbolic geometry. It is sure that the ideas of GAUSS are fundamental.

A bit disturbing is, however, the assertion appearing in the response letter of GAUSS that he was engaged by these thoughts as early as 30–35 years before. We cannot cast doubt on the truth of his words. On the other hand, if we count back this number of years from the time when the *Appendix* reached GAUSS, then we obtain essentially the interval when GAUSS and FARKAS BOLYAI stayed in Göttingen and, consequently, might have made clear their views of space in personal conversations. This possibility has led to the commentary, rather wide-spread in the world literature, that the early ideas of GAUSS were passed by FARKAS BOLYAI to JÁNOS, in other words that the *Appendix* was initiated and inspired by some early ideas of GAUSS.

We can refute this opinion by several arguments. First of all, if FARKAS BOLYAI had been well oriented about the early ideas of GAUSS, then he would not have wasted almost thirty years with the aim of proving the axiom of parallelism and cautioned his son against this dangerous problem on so many occasions. Furthermore, GAUSS himself believed at a younger age that the axiom of parallelism could be proved, as attested by one of his letters (from 1804) to FARKAS BOLYAI ([3], p. 81). We also know that at the beginning of his investigations JÁNOS BOLYAI followed the example of his father (that is, tried to prove Axiom XI); why would he have done so if he had known the Gaussian concept of non-Euclidean geometry? Probably, his father oriented him about the 18th century attempts concerning the axiom of parallelism (SACCHERI, LAMBERT, and others), but it is absolutely sure that he did not know anything of the investigations accomplished in the second decade of the 19th century (TAURINUS, SCHWEIKART, and others). The most significant thoughts of GAUSS in this direction have ripened after 1814; then, however, the two BOLYAI's had no contact with him either by correspondence or personally. Correspondence could have been the only connecting link, but for 23 years and a half (from 2nd September 1808 till 6th March 1832) no line of writing has arrived from GAUSS to the address of FARKAS BOLYAI.

We stress that our aim has not been to diminish the merits of GAUSS; rather, we wanted to support the opinion that JÁNOS BOLYAI *built up his geometrical system independently of* GAUSS.

*

In Hungary, the discovery of JÁNOS BOLYAI was received by complete indifference and incomprehension — apart from the praising words of the father. For the appreciation of the new space theory two conditions would have been necessary: the *mathematical* contents of the *Appendix* should have been understood, and then its *philosophical* implications accepted. At that stage of our mathematical life even the first condition was difficult to fulfil. In the undeveloped social circumstances of Hungary, interest in science and mathematics evolved very slowly since the feudal conditions did not require any considerable knowledge of mathematics. At the beginning of the last century many Hungarian scientists have already seen how backward we are and made efforts in various pamphlets, polemic essays and instructional programs to improve the situation. Their endeavour, however, remained mostly a meek desire; a considerable advance ensued only in the last third of the century. Thus we can take for granted that the *Appendix,* hard to follow anyway, has not been studied by anybody in our country — apart from FARKAS BOLYAI — in the first half of the 19th century. There has been also another — surprising — reason of the impassive reception of the *Appendix* at home: *its being written in Latin.* We need not think that the reading of the *Appendix* was hindered by inexperience in the Latin language; in fact, then the majority of our intelligentsia spoke Latin almost as an official language. The reason is that the *Hungarian Academy of Sciences,* which began its activity in 1830, in the first times set as its definite aim to cultivate and develop the Hungarian language, and inequitably condemned all essays published in a foreign language. According to a letter — written to the father — of the secretary of the Academy who held office at that time, János might even become a member of the Academy if his treatise appeared in Hungarian. Curiously, after a few decades the situation has turned to the opposite: essays published in Hungarian remained in several cases entirely unnoticed;

*

JÁNOS BOLYAI was very grievously affected by the lack of success. He performed his military service superficially, got along with his fellow-officers more and more badly, his physical and psychical state of health gradually deteriorated. Finally, getting tired of military life and full of desire to pursue his mathematical investigations more freely, he applied for retirement with a pension. His request has been fulfilled and on 15th June 1833 he was pensioned off without excluding the possibility of a later reinstatement. So, JÁNOS BOLYAI got home to his native country. For a short period he lived in Marosvásárhely, but soon moved to his mother's estate in Domáld (Domald; now Viişoara, Roumania). We know the reason: father and son, two men of antagonistic character could not live near each other, since scientific controversies and financial disputes disturbed their relations.

The years spent by JÁNOS BOLYAI in Domáld have been a little more quiet. Although his life was not free from financial difficulties, he had more time for mathematical meditations. According to his bequeathed manuscripts he has continued his investigations into space theory, totally isolated from the scientific world, having no mathematical literature at his disposal. According to his manuscripts, in these years he was mostly interested in two questions: first, the evaluation of the volume of the tetrahedron in hyperbolic geometry and, second, the verification of the consistency of his geometrical system.

In the letter of 6th March 1832 to FARKAS BOLYAI, also GAUSS called the attention of János to the problem of the volume of the tetrahedron ([3], p. 112), but there are indications that János has already performed such calculations before. This is almost natural as after the construction of the plane trigonometry of the Bolyai geometry the question about the volume of the tetrahedron arises immediately. GAUSS and LOBACHEVSKY have also dealt with this problem. The way followed by them is in some places very near to the argument of JÁNOS BOLYAI; nevertheless, we can say with certainty that all three of them worked independently. BOLYAI started with a tetrahedron one face of which is a right triangle and one edge of which is perpendicular to that face at the vertex of an acute angle. Let us call this tetrahedron a *normal tetrahedron*. So, each face of the normal tetrahedron is a right triangle. To find the volume, we must decompose the normal tetrahedron by some procedure into elementary parts; if we know the volumes of the latter, the required formula is furnished by a definite integral. JÁNOS BOLYAI has proposed several procedures for the partition of the normal tetrahedron, but at a certain point his investigations have broken off since the solution leads finally to an elliptic integral. His notes, difficult to follow, were published in 1901 in German and Hungarian by STÄCKEL (who, in his book, considers the question in detail; see [5], pp. 109—118).

Also just after the publication of the *Appendix*, JÁNOS BOLYAI began to study the problem whether his geometrical system involves intrinsic *contradictions*. It seems that this part of JÁNOS BOLYAI's bequest has escaped the attention of KAGAN, as in his treatises there are remarks according to which BOLYAI — in contradistinction to LOBACHEVSKY — has not dealt with the problem of consistency.

It is not our task to give the details of the witty calculations executed by JÁNOS BOLYAI in this direction (for this point cf. [5], pp. 185–187), but we indicate the starting-point. It should be noted that in BOLYAI's opinion the consistency of his geometrical system *in the plane* was verified by the fact that his trigonometrical relations formally coincided with the corresponding relations for the spherical triangle of imaginary radius. However, he deemed from the beginning that this was not a proof of the consistency of stereometry.

The aim of the method he has invented for proving the latter was to reduce the problem of consistency to the consistency of Euclidean stereometry. Still, his procedure — however ingenious — is not suitable for giving an answer of universal validity.

The essence of JÁNOS BOLYAI's plan has been the following. Given the six edges a, b, c, d, e and f of the tetrahedron (see the Figure), the

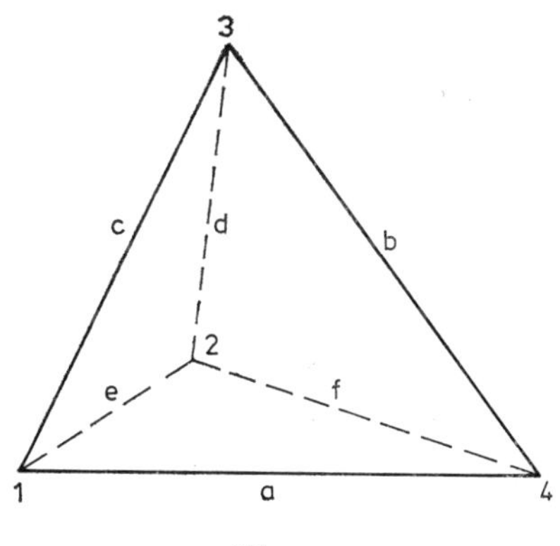

Figure

angles and *dihedral angles* can be calculated by means of the cosine law. Now the dihedral angle at edge e can be obtained in two ways, namely with the help of the *angles* at either vertex 1 or vertex 2. Clearly, relative to one and the same dihedral angle both ways must lead to the same result. Using the formulas of Euclidean trigonometry — in a fairly intricate manner — we arrive at the desired coincidence. JÁNOS BOLYAI has examined whether the situation is the same when applying the relations of hyperbolic trigonometry. *In the course of his calculations he succeeded in getting the result expected.*

Thus, in the case of the tetrahedron ("the system of four points") he has found no contradiction. Then he enlarged the system of four points by a fifth point which is not coplanar with any three of the first four. With the knowledge of *nine* edges of this body, some dihedral angles can be calculated in *three* ways. By these calculations he first did not obtain the coincidence desired. Hence he inferred that his geometrical system was inconsistent. As an effect, it has even occurred to him that so he had found — in an indirect way — the proof of Euclid's fifth postulate. However, when checking the intricate calculations he noticed that he had made a mistake and even "the system of five points" does not reveal any contradiction. According to some of his notes he wanted to proceed in this direction but he was deterred from going on by the fact that already in the case of six points the formulas were too long to look over.

Accordingly, the notes of JÁNOS BOLYAI have been discontinued so that he could never make sure the relative consistency of his geometrical system. He has shared the fortune of his remote fellow-scientist LOBACHEVSKY also in this respect. Nevertheless, it is an undying merit of both of them that they have seen clearly the basic requirement of giving consistency proofs in those sciences which are built up axiomatically, and it was on their influence that this has later become an outstanding problem of the foundations of mathematics.

*

With such investigations, among financial problems and the still alert hope of success, the days of János Bolyai passed in Domáld up to 1846. From 1834 he lived together with a girl called ROZÁLIA ORBÁN of Kibéd, but for lack of the sum *(security)* required from an officer when getting married they could — according to the register of the reformed parish of Marosvásárhely — legalize their cohabitation only on 18th May 1849. Namely at that time, as a consequence of the dethronement in Debrecen of the Lorraine–Habsburg dynasty, the decrees of the imperial court of Vienna and, in particular, the obligatory payment of a security lost their force for a short period. We know from a letter of Farkas Bolyai that after the fall of the 1848–49 war of independence the emperor did not approve of the marriage. By the way, JÁNOS BOLYAI has provided till the end of his life with fatherly affection for the two children born from this marriage.

He has remained interested, more or less intensively, in mathematical problems, but has not complied a systematic work like the *Appendix* any more. He endeavoured for several decades to build up geometry in an axiomatic way. Unfortunately, the structurally well conceived work *Raumlehre* (in German) — the preface of which was written by him in 1834 already — has remained unfinished. This is to be regretted also because the manuscript contains many valuable original ideas, especially in the field of *topology*. He has also examined questions (algebraic solution of an algebraic equation of arbitrary degree, giving a closed form to any integral, etc.) which could not bring him a success; an adequate familiarity with the literature would have saved him a lot of needless effort.

In 1837 the two BOLYAI's learned of a competition of the Jablonowski Society in Leipzig on the clarification of some questions relative to the geometric representation of complex numbers. Both the father and the son have sent in a hand-written treatise for the competition, but without success (one half of the prize was won by FERENC KEREKES, professor in Debrecen, for a much less significant work). In the case of JÁNOS BOLYAI the failure was painful, he lost his ambition for a long time again. Nevertheless, his treatise *"Responsio"* contains valuable original results. According to

JÁNOS BOLYAI, even the title of the competition was mistaken, as the problem of geometrical constructions related to the complex numbers was of secondary importance, much more important being their exact definition and the question about their role in geometry. In essence, the *Responsio* defines the complex numbers as ordered pairs of numbers, and postulates — in a somewhat circuitous manner — the rules valid for operations with these pairs. Thus it chooses basically the same way as HAMILTON did a little earlier. But it is sure that the most valuable part of the treatise is §9 — not understood by the reviewers. Namely, with a simple reference to the *Appendix,* JÁNOS BOLYAI presents there two trigonometric relations of his geometrical system, points out the notion of the hypersphere and that in his system the trigonometric relations coincide in form with the trigonometry of the sphere of radius $\frac{k}{i}$. Thus, by his brief comments he wanted to exhibit the basic importance of the imaginary unit i in his theory of space. How could the reviewers — nonentities in the history of mathematics — have understood the thoughts of JÁNOS BOLYAI? In fact, the *Appendix* may not have reached them.

*

In 1846 JÁNOS BOLYAI wound up his Domáld home and moved with his family to Marosvásárhely. So the father and the son lived again near each other but not without frictions. There was one more event which broke the monotony of their life: through a newspaper article of a Transylvanian student who had met GAUSS, in 1844 FARKAS BOLYAI became aware of LOBACHEVSKY's book *Geometrische Untersuchungen zur Theorie der Parallellinien* (Berlin, 1840). However, they could get the book only after several years, in 1848. Both of them have studied the work thoroughly and with curiosity; also, they added comments to it. From their comments it turns out that they have seen the significance of the work clearly and understood its contents. FARKAS BOLYAI in his last mathematical work *Kurzer Grundriss* (in German; Marosvásárhely, 1851; published in full also by STÄCKEL; see [5], Volume II, pp. 119–179) devotes several pages (pp. 175–179) to a comparison between some results of LOBACHEVSKY and JÁNOS, and completes a defective proof of the scientist "from Kazan".

Unfortunately, the comments of JÁNOS BOLYAI on LOBACHEVSKY's work have been left behind only in manuscript and appeared in print much later (see e.g. [5], Volume I, pp. 140–160). These comments give evidence that JÁNOS BOLYAI first received LOBACHEVSKY's work distrustfully, but as he went on with the study of it, his distrust gradually changed to appreciation and, what is more, admiration. Objective judgement has been characteristic of JÁNOS BOLYAI, especially when ranking mathematical results. His human greatness is documented by the following sentence, alluding to LOBACHEVSKY, of his manuscripts left behind: "... I am sharing the finder's merit with pleasure" ([12], p. 77). And we here discover one more argument which shows that the thoughts of JÁNOS BOLYAI could not be suggested by GAUSS: in fact, why would

Bolyai not make a similar statement about the "Colossus of Göttingen" appreciated by him so much if he had known of the ideas of Gauss concerning space theory? On the other hand, as a consequence of the reply of Gauss on receiving the *Appendix,* he notes at several places that Gauss has also been engaged in the problem of the possibility of non-Euclidean geometries, but these remarks are always accompanied by the term "allegedly". This indicates that he has not been acquainted with the results of Gauss in the field of hyperbolic geometry.

*

At the end of 1852 János Bolyai broke off relations with Rozália Orbán, and this has removed the main reason of the quarrel between father and son. Mistrust and disagreement gradually turned into mutual affection and the full recognition of the mathematical activity of the other. So in 1856 when the father, the old professor was payed the last honours by the bell-ringing of the college and by the inhabitants of the town, János Bolyai has been left orphaned in two senses: he lost his father and the only fellow-scientist who had understood him.

János Bolyai in the last part of his life has no more dealt with mathematical problems regularly; he has rather worked on a book having a mostly philosophical subject and aiming at the happiness of mankind. According to the bequeathed first draft of *Salvation Theory* ([12], pp. 177–255), he kept his clear logic to the last. His — mainly *linguistical* and *progressive sociological* — thoughts given in outline promised a truly valuable work, but in those years he had already not enough physical and psychical strength for systematic labour.

Fast decline of the body was soon followed by the final break-down: on 27th January 1860 one of the greatest figures of Hungarian science died — perhaps in pneumonia — without the tears and sympathy of others.

No authentic picture of him has remained. From his manuscripts left behind we know that the only oil-painting made of him and representing him in a second-lieutenant's uniform was destroyed by himself in a moment of despair. That is why we cannot fulfil the numerous requests received in this direction from home and abroad. The picture published of him at several places (even on the stamp issued by the Hungarian and the Roumanian post) is *non-authentic.*

The funeral was as tragic as distressing had been his whole career, especially the last part of his lonely life: the funeral procession consisted, apart from the two officers dispatched by the authorities, of only three civilians. Even the exact place of his grave was unknown for a long time. After a careful investigation, on 11th June 1911, father and son who had been lying in different graves were disinterred in compliance with official regulations, and their mortal remains were put in a common grave. From that time on, the two scientists who quarrelled so much when alive are resting under one and the same burial-mound in the reformed cemetery of Marosvásárhely.

*

In the foregoing we have mentioned several times that the mathematical activity of the two Bolyai's — while they lived — did not bring about any considerable reaction either in Hungary or abroad. To be more precise: there have appeared some brief criticisms, in Hungarian, of the books of FARKAS BOLYAI which were very unfavourable, but the name of JÁNOS BOLYAI is not to be found anywhere in print. So both of them died with the painful feeling of incomprehension, without recognition and success.

It is less well known who were the first to understand the significance of their work, and after what antecedents our two scientists have occupied their due place in the history of mathematics.

The pioneers of "discovering" the two BOLYAI's were not Hungarian; it was under the influence of initiative from abroad that our scientific life realized its omissions and duties in this respect.

The first writing that mentioned the name BOLYAI — mainly in connection with the father — in a way accessible to a wider public was an essay devoted to the memory of GAUSS: SARTORIUS VON WALTERSHAUSEN, when arranging the bequest of GAUSS and hearing of the correspondence between FARKAS BOLYAI and GAUSS, asked FARKAS for the letters of GAUSS. He has taken into consideration also this material when he compiled his obituary of GAUSS (1856) containing also some biographic data of the two BOLYAI's.

The German university professor RICHARD BALTZER (1818–1887) has been the first to treat in effect a few mathematical results of the two BOLYAI's in his two-volume work *Die Elemente der Mathematik* (1860 and 1862). BALTZER's activity is important as his book has been published several times. The work mentions the definition of parallelism given by JÁNOS BOLYAI, and the theorem which says that parallelism defined in this way is also a transitive relation.

One of the most significant dates in the history of the Bolyai geometry is the year 1867 when the Appendix appeared *in French,* translated by G. J. HOÜEL (1823–1886), professor at the university of Bordeaux. HOÜEL learned of the activity of JÁNOS BOLYAI and LOBACHEVSKY from the work of BALTZER mentioned above and, observing the importance of the matter with an amazing perspicacity, published first LOBACHEVSKY's book *Geometrische Untersuchungen zur Theorie der Parallellinien* and then soon — in the same year — the *Appendix*, both in French. As a consequence, the treatise of JÁNOS BOLYAI has become more accessible to those interested in it. It was at this stage that one of the most persevering and most effective explorers of the Bolyai problem, not a mathematician by profession, the Hungarian architect FERENC SCHMIDT (1827–1901) joined in the work. His father had known JÁNOS BOLYAI at Temesvár personally and, as a result of his recollections as well as own assembling, SCHMIDT could compile the biographical data of the two BOLYAI's in more detail. Accidentally, HOÜEL has heard of SCHMIDT's activity and asked him to write down his results. In 1867 Schmidt's treatise appeared in French and German and, on insistence of the Italian historian of mathematics BONCONPAGNI (1821–1894), in the next year even *in Italian*. In 1868 also the *Appendix* was published in *Italian*.

Hungarian scientific life has not remained uninfluenced by these writings; one can say that the conscience of our experts has been waked by them. JENŐ HUNYADI (1838–1889), professor of mathematics at the Budapest Institute of Technology, proposed on a session of the Hungarian Academy of Sciences in 1868 that a commission should examine the bequest of the two BOLYAI's and make a motion towards publishing their more significant results. Ignorant of the Hungarian initiative, BONCONPAGNI on two occasions (1869 and 1871) called the attention of our scientific circles to the importance of scrutinizing the BOLYAI bequest.

Although these events have aroused interest in the work of JÁNOS BOLYAI, actual recognition has arrived only after a longer time. Neither the activity of LOBACHEVSKY had much better luck. Even RIEMANN's famous *Habilitationsschrift* (1854) was not crowned with immediate success which only partly originated in the fact that it appeared in print much later, in 1867.

In any case, the indifferent, sometimes definitely hostile reception of the first non-Euclidean geometries was somewhat modified by BELTRAMI's dissertation published in 1868 according to which the theorems of hyperbolic geometry are locally true on the pseudosphere. This, however, did not bring to an end the obstinate resistance against the new geometries, though from that time on the opposers were mostly *philosophers,* not mathematicians. The leading personality of this struggle was R. H. LOTZE (1817–1881), professor of philosophy at Göttingen. He and his school have taken for nonsense all geometries different from the Euclidean. Among their arguments there appeared the theorems of non-Euclidean geometry unusual for our intuition, and the — explicitly mathematical — notion "measure of curvature" introduced by GAUSS. The majority of philosophers could not separate from KANT's idealistic concept of space and regarded the "curvature" of non-Euclidean spaces as a foolish mystification. As a matter of fact, several writings from the sixties and seventies of the last century include statements according to which the non-Euclidean geometries are unacceptable and contradict sound geometric intuition.

The progressive Austrian mathematician JOHANNES FRISCHAUF (1837–1924), professor at the university of Graz, was one of the victims of this hostile attitude. In connection with JÁNOS BOLYAI his personage deserves particular attention since in the academic year 1871–72 he already gave a course on non-Euclidean geometries, relying mainly on the *Appendix. This course was the first detailed exposition of the Appendix,* significant also because it has appeared in print (*Absolute Geometrie nach Johann Bolyai*, Leipzig, 1872). This booklet has long been the only work to build up non-Euclidean geometry, relying definitely on the *Appendix,* by an elementary synthetic method. We know from the preface of the book that FRISCHAUF has originally planned a publication of the *Appendix* with comments, but was diverted from his purpose by the information that GYULA KÖNIG (1849–1913), professor at the Budapest Institute of Technology, was already working on such a publication (because of other engagements, KÖNIG has relinquished the project). FRISCHAUF's book does not treat those problems of construction which appear in §§ 34–43 of the *Appendix.*

As far as we know, BOLYAI's system of geometry won the name "absolute geometry" as a consequence of the title of FRISCHAUF's book. Although the term "absolutely true" occurs in the full title of the *Appendix* and at many places of various writings of the two BOLYAI's its use is always *attributive*. As a *title*, JÁNOS BOLYAI himself applied one of the expressions "space theory", "Raumlehre", "Scientia Spatii", or denoted his system of geometry by the letter **S** (in contrast to Euclidean geometry which he denoted by Σ).

Another book of FRISCHAUF entitled *Elemente der absoluten Geometrie* (Leipzig, 1876) is a little more comprehensive and detailed than the first one but has similar contents. The *Appendix* became well known both in Hungary and abroad, first of all, through the works of FRISCHAUF. His role should be stressed also because, owing to his up-to-date lectures and fearless behaviour, he had to endure violent attacks of the Austrian educational organs and some scientific circles.

In rendering non-Euclidean geometries recognized and verifying their relative consistency, distinguished services have been made by FELIX KLEIN. The circumstances from which his studies in this direction took their origin are described in his own work on the history of mathematics ([7], pp. 151–152). We know from there that he learned about the results of LOBACHEVSKY and JÁNOS BOLYAI in 1869, even then second-hand, through exposition by STOLZ. In the course of endless disputes in Berlin — in the seminar of WEIERSTRASS — he arrived at the conviction that non-Euclidean geometries could be treated also as special chapters of projective geometry. He published his ideas in 1871–72, in particular the construction that is called nowadays the Cayley–Klein model. It is little known that simultaneously with FELIX KLEIN also GYULA KÖNIG proposed a model for representing non-Euclidean geometries [1], but his paper — perhaps as a consequence of the sketchy composition — has remained unnoticed.

In the meantime the "Bolyai Commission" (members: JÁNOS ÁRMIN VÉSZ, GYULA KÖNIG, JENŐ HUNYADY and FERENC SCHMIDT) organized by the Hungarian Academy of Sciences completed its report, marked out those parts of the bequest of the two BOLYAI's which ought to be republished. Implementing the project, however, has still demanded a long time. But the years of delay have not passed fruitlessly: indeed, while the bequest was processed, one has discovered results of the Bolyai's, one after the other, that offered possibilities of generalization and further investigation. *Thus in the last two decades of the 19th century a scientific literature relying on the results, and fed by writings, of the two Bolyai's has evolved in this country.*

In the following we mention only some of the more important items, related to JÁNOS BOLYAI, of this literature. The introduction was supplied by a lecture on the *Appendix* delivered by the university professor MÓR RÉTHY (1848–1925) in 1874; it soon appeared in print. RÉTHY's treatise has been a work of great importance in making known and popularizing the Bolyai geometry. Réthy's aim was to create desire for studying the *Appendix*. To make this easier, he presented more comprehensible proofs of some theorems of the *Appendix*. MÓR RÉTHY was the first to elaborate the beautiful problems of hyperbolic construction appearing in the *Appendix*. He advocated the principle that mathematical research in Hungary, first of all, ought to start from, and rely on, the activity of the two BOLYAI's.

The work of RÉTHY at the university of Kolozsvár was continued very effectively by GYULA VÁLYI (1855–1913) who, beginning with the second semester of the academic year 1891–92, gave a special course "*On the Appendix by János Bolyai*" several times. Unfortunately, this extremely well considered and accurately organized series of lectures has not appeared in print; nevertheless, a few duplicate copies of it have been preserved. According to these copies, VÁLYI devoted about one third of his course to the exposition of historical antecedents and then commented on the *Appendix*, proceeding in the order of the sections. He completed BOLYAI's proofs, mitigated the terseness of style by inserting explanatory passages and, here and there, compared the results of JÁNOS BOLYAI with those of LOBACHEVSKY. Undoubtedly, this commentary had an effect on subsequent authors (LAJOS DÁVID).

It was an important stage of the recognition of JÁNOS BOLYAI when in the nineties of the last century G. B. HALSTED (1853–1922), professor of mathematics at the University of Texas, Austin, joined in the popularization of non-Euclidean geometries. First the treatise of LOBACHEVSKY appeared *in English* translation made by him, whereas in 1891 the first English edition of the *Appendix* — followed by three further editions within three years — was published. The particular importance of HALSTED's activity consists in that, being an entirely impartial person enthusiastic about the cause, he considered it his task to make people realize that LOBACHEVSKY and JÁNOS BOLYAI should be held in the same esteem. In fact, up to that time the treatises of LOBACHEVSKY happened to be published more often and commented upon in more detail. The Russians gave more care to gaining recognition for their outstanding scientist. Thus, for instance, they have erected a statue of LOBACHEVSKY in Kazan, and established a prize of considerable amount to immortalize his memory. In 1895 HALSTED gave an account of these events in a New York journal of science, alluding almost reproachfully to the negligence of the scientific circles of Hungary. But HALSTED's respect for JÁNOS BOLYAI has meant even more: in the summer of 1896 he went on a pilgrimage to the grave of JÁNOS BOLYAI. He used that opportunity to ask JÁNOS BEDŐHÁZI (1853–1915), professor at the college of Marosvásárhely, to write a comprehensive monograph on the two BOLYAI's. The work of Bedőházi [2] is useful mainly with regard to biographical data; it tells us comparatively little about the mathematical activity of the two BOLYAI's.

In the last years of the 19th century, French scientists under the chairmanship of POINCARÉ have complied an extensive bibliography of mathematics. That chapter of the publication which enumerates works on non-Euclidean geometries received — on intercession of the Hungarian scientists — the heading "Bolyai–Lobachevsky geometry". From that time it has become standard in mathematical literature all over the world to specify the first non-Euclidean geometries by the joint use of the names of both scientists.

After a protracted preparation lingering through several decades, the second edition of the *Tentamen* was published in 1897 and 1904. In its second volume, on pp. 359–394, also the *Appendix* is included. On the other hand, the *Appendix* appeared *in Hungarian* already in 1897, at once in two translations. One has been the work of IGNÁC RADOS, the other — the better one as a translation — that of JÓZSEF SUTÁK. SUTÁK has added a commentary to his translation, but several of his comments are wrong.

About the turn of the century the enthusiasm which has been destined for disclosing the life and activity of the two BOLYAI's to the whole world ran actually high. In 1899, after careful preparatory work, the correspondence of FARKAS BOLYAI and GAUSS was published [3]. While this publication was being prepared, the German mathematician PAUL STÄCKEL (1862–1919) who had a good knowledge of the activity of GAUSS in the field of geometry but was also enthusiastic about the results of the two BOLYAI's joined in the work. During his investigations, in order to learn the Bolyai problem more thoroughly, he even visited Marosvásárhely and, undeterred by linguistic difficulties, went through the bequest (with the help of JÓZSEF KÜRSCHÁK). STÄCKEL published his research in a whole series of essays in Hungarian and German. Several mathematical results of JÁNOS BOLYAI — not to be found in the *Appendix* — have been revealed then. The two-volume work [5] written by him — the second volume contains translations from the bequest of the two BOLYAI's — has remained, in spite of minor inaccuracies, the standard work which treats the activity of the two BOLYAI's the most profoundly.

After antecedents of this kind the hundredth anniversary of JÁNOS BOLYAI's birth arrived. On this occasion a decoratively got-up collection of essays was published [4]. At the same time, the Hungarian Academy of Sciences — taking the Lobachevsky Prize as an example — established a "Bolyai Prize". In agreement with the foundation document, the prize was awarded in 1905 for the first time, and the intention was to award it subsequently in every fifth year "for outstanding mathematical investigations published anywhere and in any language". Unfortunately, the good initiative could be realized only two times: in 1905 the prize was awarded to POINCARÉ, and in 1910 to HILBERT. During World War I, however, international scientific relations became irregular, our currency was loosing in value, and the undertaking was discontinued.

*

This is a brief sketch of the first period of discovering the two BOLYAI's. Also in our century there have always been Hungarian and foreign scientists who revived the memory of the BOLYAI's in a monograph or obituary, but really abundant results we owe most of all to the various *Bolyai anniversaries* held in the second half of the 20th century. The number of tanslations of the *Appendix* has also increased. Special mention is deserved by the *Russian* version (Moscow–Leningrad, 1950) translated and commented by V. F. KAGAN. Moreover, KAGAN compared results in non-Euclidean geometry due to GAUSS, BOLYAI and LOBACHEVSKY in a long treatise of rather high level [9]. Discussion of recent literature on the subject would go beyond the scope of this Supplement.

LITERATURE

[1] König, Julius: *Über eine reale Abbildung der s. g. Nicht-Euklidischen Geometrie,* Götting. Nachrichten, 1872, 157–164.

[2] Bedőházi, János: *The Two Bolyai's,* Marosvásárhely, 1897 (in Hungarian).

[3] *Briefwechsel zwischen Carl Friedrich Gauss und Wolfgang Bolyai,* Leipzig, 1899.

[4] *Ioannis Bolyai in Memoriam,* Claudiopoli, 1902 (a collection of essays).

[5] Stäckel, Paul: *Wolfgang und Johann Bolyais geometrische Untersuchungen,* I–II. Leipzig—Berlin, 1913.

[6] Dávid, Lajos: *The Life and Work of the Two Bolyai's,* Budapest, 1923 (in Hungarian).

[7] Klein, Felix: *Vorlesungen über die Entwicklung der Mathematik im 19.,* Jahrhundert. Berlin, 1926.

[8] *Briefwechsel zwischen Carl Friedrich Gauss und Christian Ludwig Gerling,* Berlin, 1927.

[9] Kagan, V. F.: *The Construction of Non-Euclidean Geometry by Lobachevsky, Gauss and Bolyai,* Proc. Inst. History of Science, 1948, II, 323–399 (in Russian).

[10] Alexits, György: *János Bolyai,* Budapest, 1952 (in Hungarian).

[11] *The Life and Work of János Bolyai,* Bucharest, 1953 (a collection of essays; in Hungarian).

[12] Benkő, Samu: *The Confessions of János Bolyai,* Bucharest, 1968 (in Hungarian).

SUPPLEMENTARY LITERATURE

To supplement the bibliographic references appearing in the book, but without pretending to completeness, we call the attention of the reader to the following works.

BACHMANN, F.: *Aufbau der Geometrie aus dem Spiegelungsbegriff,* Berlin—Göttingen—Heidelberg, 1959.

BALDUS, R.: *Nichteuklidische Geometrie,* Sammlung Göschen, 1964.

BLUMENTHAL, L. M.: *A Modern View of Geometry,* Freeman and Company, 1961.

COXETER, H. S. M.: *Non Euclidean Geometry,* Toronto, 1957.

GREENBERG, M. J.: *Euclidean and Non-Euclidean Geometries,* Development and History. San Francisco, 1974.

KÁRTESZI, F.: *Introduction to Finite Geometries,* Akadémiai Kiadó, Budapest, 1975.

KERÉKJÁRTÓ, B. V.: *Les fondaments de la géometrie,* I—II. Budapest, 1955 and 1966.

LENZ, H.: *Nichteuklidische Geometrie,* Mannheim, 1967.

SZÁSZ, P.: *Introduction to the Bolyai–Lobachevsky Geometry,* Akadémiai Kiadó, Budapest, 1973 (in Hungarian).